FAST SPACE

LEVERAGING ULTRA LOW-COST SPACE ACCESS FOR 21ST CENTURY CHALLENGES

AIR UNIVERSITY

NIMBLE BOOKS LLC: THE AI LAB FOR BOOK-LOVERS
~ FRED ZIMMERMAN, EDITOR ~

Humans and AI making books richer, more diverse, and more surprising.

Publishing Information

(c) 2023 Nimble Books LLC
ISBN: 978-1-60888-197-0

AI-generated Bibliographic Keyword Phrases

US Air Force; strategic advantage in space; space architecture vulnerability; competition; Fast Space approach; ecosystem of concepts, capabilities, and industrial partnerships; speed as defining attribute of advantage in space; sortie-on-demand launch capability; economically viable business cases; high launch rates; lower costs; rapid turn-around; higher reliability; reducing costs through reusability and increased flight rates; partnering with US commercial firms pursuing ultra-low-cost access to space; purpose-built organization for managing partnerships;

Publisher's Notes

This report played a crucial role in the decisions involved in creating United States Space Force and remains an essential guide to thinking about the strategic value of ultra-low-cost launch.

This annotated edition illustrates the capabilities of the AI Lab for Book-Lovers to add context and ease-of-use to manuscripts. It includes five types of abstracts, building from simplest to more complex: TLDR (one word), ELI5, TLDR (vanilla), Scientific Style, and Action Items; three essays to increase viewpoint diversity: Grounds for Dissent; Red Team Critique; and MAGA Perspective; and Notable Passages and Nutshell Summaries for each page.

Fred Zimmerman
Ann Arbor, Michigan, USA

AI-GENERATED ANNOTATIONS

Publishing Information ii
AI-generated Keyword Phrases ii
Abstracts iv
    TL;DR (one word) iv
    Explain It To Me Like I'm Five Years Old iv
    TL;DR (vanilla) iv
    Scientific Style iv
    Action Items .... **Error! Bookmark not defined.**
Viewpoints .... **Error! Bookmark not defined.**
    Grounds for Dissent vi
    Red Team Critique viii
    MAGA Perspective x
Page-by-Page Summaries xi
Notable Passages xvii

ABSTRACTS

TL;DR (ONE WORD)

Space.

EXPLAIN IT TO ME LIKE I'M FIVE YEARS OLD

This document is about how the US Air Force can use new ideas and technology to be better in space. It talks about how the current things we use in space are not very safe and can be easily attacked by other countries. So, the US needs to find new ways to be better in space. The document suggests a "Fast Space" plan, which means being really fast in space. This plan includes being able to launch spaceships whenever we want, making it cheaper and faster to go to space

TL;DR (VANILLA)

This document discusses the need for the US Air Force to leverage new concepts and technologies in space to gain strategic advantage. It proposes a "Fast Space" approach that prioritizes speed, lower costs, and higher reliability. The document recommends partnering with commercial firms and shaping the regulatory environment to achieve these goals.

SCIENTIFIC STYLE

This document presents a proposal for the US Air Force to gain strategic advantage in space by leveraging new concepts and technologies. It highlights the vulnerability of the current space architecture and the need for the US to regain advantage amidst growing competition. The proposed "Fast Space" approach prioritizes speed as the defining attribute of advantage in space and includes sortie-on-demand launch capability, economically viable business cases, high launch rates, lower costs, rapid turn-around, and higher reliability. The document emphasizes reducing costs through reusability and increased flight rates, recommending partnerships with US commercial firms pursuing ultra-low-cost access to space. It also suggests creating a purpose-built organization to manage these partnerships and shaping the interagency environment to ease regulatory burdens. Integrating high launch rate rapid-turnaround approaches into the joint requirements and acquisition process is also recommended. Overall, this document highlights the need for the US to capitalize on the changing landscape of space exploration and exploitation.

RECOMMENDED FOLLOW-UP ACTIONS

Develop and implement a sortie-on-demand launch capability to increase flexibility and responsiveness in space operations.

Foster partnerships with US commercial firms working on ultra-low-cost access to space to leverage their expertise and resources.

Create a dedicated organization within the US Air Force to manage these partnerships and ensure effective collaboration.

Advocate for regulatory changes that facilitate rapid-turnaround approaches and reduce bureaucratic burdens.

Integrate high launch rate rapid-turnaround approaches into joint requirements and acquisition processes to prioritize speed as a defining attribute of advantage in space.

Invest in research and development of new technologies that can enhance the US Air Force's access to space.

GROUNDS FOR DISSENT

Ethical concerns: A member of the organization might have principled reasons to dissent from this report based on ethical concerns. They may argue that prioritizing speed and advantage in space could lead to a militarization of space, potentially escalating conflicts and endangering peaceful exploration and cooperation. They could believe that the focus should be on international collaboration and ensuring the peaceful use of outer space, rather than seeking strategic advantage.

Environmental impact: Another reason for dissent could be based on substantive concerns regarding the environmental impact of the proposed "Fast Space" approach. The member might argue that rapid launch rates and increased flight activity could contribute to space debris, which poses significant risks to existing satellites and future space missions. They may advocate for a more sustainable approach that considers the long-term consequences for both Earth's orbit and future space exploration.

Resource allocation: A member might dissent due to concerns about resource allocation. They may argue that investing heavily in pursuing ultra-low-cost access to space through partnerships with commercial firms could divert resources away from other crucial areas such as education, healthcare, or addressing climate change on Earth. They may believe that priorities should be reassessed to ensure a balanced allocation of resources across various societal needs.

Economic inequality: There could also be substantive reasons for dissent related to economic inequality. The member might assert that promoting economically viable business cases and focusing on reducing costs through reusability primarily benefits wealthier nations or corporations with greater financial capabilities, exacerbating existing disparities between countries and widening the gap between developed and developing nations' access to space technology. They might call for a more equitable approach that considers the equal distribution of opportunities in the realm of space exploration.

Lack of public input: Dissent might arise from concerns about democratic principles and lack of public input in shaping these strategies. The member might argue that decisions about the direction of national space policy should involve broader public participation rather than being solely determined by the military or government agencies. They may advocate for increased transparency and dialogue to ensure that diverse perspectives are considered in shaping space exploration strategies.

Overall, these dissenting views highlight potential concerns related to ethics, the environment, resource allocation, economic inequality, and democratic participation. Members of the organization responsible for the document might have principled, substantive reasons to dissent based on their assessment of these issues and their belief in alternative approaches to space exploration and exploitation.

Red Team Critique

The document provides a comprehensive overview of the potential strategic advantages that can be gained by the US Air Force in space. However, there are several areas where further analysis and clarification are needed.

Firstly, while the document acknowledges the increasing vulnerability of the current space architecture, it does not provide sufficient evidence or examples to support this claim. A more detailed assessment of the specific vulnerabilities and their potential impact on national security would strengthen the document's argument.

Secondly, although the "Fast Space" approach is described as an ecosystem involving concepts, capabilities, and industrial partnerships prioritizing speed as a defining attribute of advantage in space, there is limited discussion on how these principles will be implemented. The document should provide more detailed explanations and examples of how sortie-on-demand launch capability, high launch rates, lower costs, rapid turn-around times, and higher reliability will be achieved.

Furthermore, while reducing costs through reusability and increased flight rates is recommended as a key objective for gaining advantage in space, no mention is made about how these goals will be achieved or what steps need to be taken to implement reusable technologies. This oversight weakens the viability of this proposal.

Additionally, while partnering with US commercial firms pursuing ultra-low-cost access to space is mentioned as a recommendation for leveraging new technologies and capabilities effectively; there needs to be an assessment of specific criteria for selecting these partners and ensuring that they align with national security objectives. Without proper evaluation processes outlined in this regard could lead to ineffective partnerships or compromising sensitive information.

Moreover,, creating a purpose-built organization to manage these partnerships requires more explanation regarding its structure and function within existing government frameworks. An alternative consideration could involve leveraging existing organizations or establishing clear guidelines for collaboration between different stakeholders involved in managing such partnerships.

Lastly , integrating high launch rate rapid-turnaround approaches into joint requirements development process also lacks details on how it can practically take place within bureaucratic acquisition procedures without adversely impacting quality assurance standards or compromises cost efficiency and safety precautions.

In conclusion, while the document provides a compelling argument for the US Air Force to seize opportunities in space exploration and exploitation, further analysis and elaboration is needed to address vulnerabilities and effectively implement proposed strategies. The recommendations should be supported by more comprehensive evidence, detailed implementation plans, evaluation criteria for partnerships, clarification on organizational structure for managing partnerships, and a clear understanding of integration into existing bureaucratic acquisition processes.

MAGA Perspective

This document is just another example of the globalist agenda pushing for a weaker America. It's clear that the authors of this document are more interested in appeasing multinational corporations than in protecting our national interests. By prioritizing speed and cost-efficiency, they are sacrificing the quality and security of our space capabilities.

The idea of partnering with commercial firms to access space at ultra-low costs is concerning. This will only lead to outsourcing important defense technologies to foreign entities, jeopardizing our national security. We should not be relying on profit-driven businesses to safeguard our strategic advantage in space. Instead, we should invest in our own capabilities and maintain control over these critical assets.

Furthermore, the suggestion to shape the interagency environment and ease regulatory burdens is nothing but an attempt to undermine our own government's oversight and accountability. We need strong regulations to ensure that the pursuit of speed and lower costs does not compromise safety or integrity.

This document also fails to address the potential risks associated with increased launch rates and rapid turn-arounds. By rushing into space missions without proper testing and evaluation, we are putting both astronauts' lives and valuable equipment at risk. The focus on high launch rates is a shortsighted approach that disregards the importance of thorough preparation and caution.

Ultimately, this document is just another attempt by the establishment elites to weaken America for their own gain. We must reject this fast-paced, cost-cutting mentality and instead prioritize robustness, security, and American exceptionalism in our space endeavors.

Page-by-Page Summaries[1]

0   This page is a white paper discussing the use of ultra-low-cost space access to address 21st-century challenges.

1   This page is a disclaimer stating that the views expressed in an academic research paper do not reflect the official policy or position of the US Government or the Department of Defense. The paper is not copyrighted and is considered property of the US Government.

2   The page provides a table of contents for a report on the changing geo-strategic landscape and risks to international stability due to increased competition in space. It discusses the concept of fast space as a strategy for US advantage, obstacles to this approach, and the benefits it could bring to the warfighter and national security.

3   This page discusses the need for the United States Air Force to adapt to changing global dynamics and explores the potential of private sector partnerships in reducing launch costs and transforming power projection capabilities.

4   This page discusses the concept of Fast Space, which is a new approach to space that focuses on speed as the defining attribute of advantage. It explores how leveraging new concepts and technologies can help the US Air Force regain strategic stability in the face of increasing competition in space.

5   Air University discusses the potential benefits of a Fast Space architecture, including affordable payloads, joint payloads for immediate effect, robust communications systems, interoperable ISR systems, global effects delivery, empowering allies, and a deployable launch-on-demand system. The attractiveness of this architecture is attributed to changing conditions such as private sector investment, emerging technological approaches, government use of Other Transaction Authorities (OTAs), and advances in manufacturing and engineering collaboration systems.

6   The page discusses the importance of reducing costs and increasing flight rates in order to achieve the benefits of a Fast Space architecture. It recommends that the Air Force partner with private space industry leaders through Other Transaction Authorities (OTAs) to capitalize on their expertise and innovative cultures.

7   The page recommends partnering with commercial firms for low-cost access to space, creating a new organization to manage these partnerships, shaping the regulatory environment, and integrating high launch rate approaches into the acquisition process.

8   This page discusses the potential for major breakthroughs in national security and industry through the affordability of space access. It proposes a "Fast Space" concept that utilizes reusable launch vehicles to enable rapid applications for the Joint Force and coalition partners.

9   Increased competition in space poses threats to US power projection and coalition partnership. The US is dependent on space for power projection, but its current space architecture is vulnerable. Nuclear deterrence is not effective against all states or groups, and US satellites are at risk of attack. Powerful states like China and Russia have invested in military capabilities that counter US power. The strategic environment in space is changing rapidly with the involvement of private industry and other strong states.

10 Competition in space has returned and the US military's advantages in space are no longer assured. The study proposes a Fast Space approach, focusing on speed as

---

[1] Page numbering as overprinted on body of original document.

the defining attribute of advantage in space. This includes sortie-on-demand launch capability, affordable payloads, commercial development, robust communications, and a disaggregated network of interoperable ISR systems.

11   Fast Space architecture offers the ability to deliver various effects worldwide, empower allies, and strengthen strategic stability. It enhances nuclear deterrence and conventional capability while accepting vulnerability and building resiliency.

12   Fast Space is a concept that aims to develop smaller, cheaper satellites and increase space launch capabilities to shape the strategic environment. Past proposals have failed due to lack of military need, high costs, technical infeasibility, and a rigid acquisition process. However, changing conditions such as private sector investment present new opportunities for Fast Space architecture.

13   Private investment and maturing technologies in the space industry are driving down launch costs, making space exploration more accessible. Other Transaction Authorities (OTAs) have proven effective in breaking traditional cost equations and aligning incentives between government and industry. The time is ripe for partnerships through OTAs to make the Fast Space vision a reality.

14   The market for ultra-low-cost access to space (ULCATS) has not yet developed due to the high risk and cost of developing commercial reusable launch vehicles (RLVs). However, an infusion of government investment and commitment could jump-start this cycle and lead to higher flight rates, decreasing costs, and increased demand.

15   The page discusses the benefits of using Other Transaction Authority (OTA) to fund partnerships between the US government and private space industry leaders. It argues that this partnership can stimulate a virtuous cycle and support a Fast Space strategy, providing new options for the Joint Force and national command authorities.

16   The page discusses the benefits of the Fast Space architecture and the overlap between developments in the commercial space sector and national security needs. It emphasizes the importance of partnering with commercial industry to fully realize the benefits available to the warfighter. The private sector's advancements in reducing launch costs and creating small satellite constellations have significant potential for contributing to Air Force missions.

17   The Air Force could partner with industry to create affordable global C2 and ISR constellations, leveraging private sector investments. This would enhance communication pathways and situational awareness for the joint team and nation, while also providing potential benefits such as ad hoc command and control structures and a more resilient global positioning system.

18   Fast Space, a proposed investment in space-based assets, aligns with the Air Force's vision for the future and can support various core missions. It offers capabilities such as high-speed transit through space, timely launch support, mobility operations, and revolutionized battle management command. Additionally, it prepares for future innovations in the space industry and the potential for large numbers of private citizens living and working in space.

19   Fast Space, a proposed technology, will enable the Air Force to operate in air, space, and cyberspace with domain superiority. It will enhance decision-making, global mobility, precision strike capabilities, and leverage commercial technologies. Using existing reusable first stages is more cost-effective than developing new ones. ULCATS has the potential to transform USAF C2 and ISR.

20   Commercial space industry developments align with USAF needs, offering affordable solutions for global C2, ISR, and communications. Commercial reusable

21 vehicles could provide prompt global strike capability, while theater pop-up missions for suborbital RLVs offer potential benefits. ULCATS systems could act as a stabilizing deterrent to war by rapidly reconstituting satellites in orbit. Co-investing with industry can influence plans for mutual benefit and mitigate risks of industry developing systems without USG involvement.
21 The page discusses the need for purpose-built vehicles in the aviation industry and the lack of JROC-validated requirements for operationally responsive launch. It highlights the potential benefits of fully reusable, aircraft-like operability launch vehicles with standardized interfaces and significant payload capacity.
22 The lack of clearly articulated and validated requirements for operationally responsive launch is due to uncertainties in the threat environment. The US Air Force needs to provide leadership in articulating emerging requirements and bringing together capabilities provided by private firms.
23 US commercial industry believes that the development of two-stage-to-orbit (TSTO) reusable launch vehicles (RLVs) is the path to achieving Ultra Long-duration Capabilities and Augmented Terrestrial Survivability (ULCATS). Multiple US companies are investing in TSTO RLVs, which are technically feasible and more economically affordable than other methods. The primary barrier to RLV development is the lack of proven market-based demand to justify private investment. Industry supports partnerships using risk-sharing Other
24 Developing fully reusable launch vehicles is a difficult technical challenge, but the benefits are worth the risk. New technologies and input from system integrators will be crucial in achieving higher levels of reusability and reducing costs. Lowering launch costs by 10X requires jump-starting industry competition with commercial RLVs to open up new markets.
25 The development of multiple reusable launch vehicles could lower prices by 3X, but it may not be enough to achieve a 10X reduction in launch costs. Key cost drivers are flight rate, reuse rate, and labor intensity of operations. Labor intensity is often ignored but is crucial for cost reduction. Redesigning how the vehicles are operated can lead to larger cost reductions and drive demand for more flights.
26 Advances in reusability and competition are necessary for significant price reductions in the space launch industry. Reusability can lead to cost savings, increased availability, and higher flight rates, creating a virtuous cycle of innovation and market growth. The early decades of aviation provide an example of how this cycle can be established through entrepreneurial innovation, competition, and government support.
27 The page discusses the impact of technological advancements in the aviation industry, particularly the introduction of new planes like the Northrop Alpha, Boeing 247, and Douglas DC-2. It also highlights the significance of the DC-3 in revolutionizing commercial aviation and how reusability is crucial for achieving ultra-low-cost access to space.
28 Big LEO constellations of small satellites for global broadband communications face challenges due to high launch costs and limited customer demand. Current launch demand is insufficient to support a significant reduction in launch costs, but new markets could potentially increase flight rates. Early purchases of commercial services by USG entities may be necessary to attract private capital. The relationship between ULCATS and big LEO constellations is synergistic, as ULCATS improves the business case for large constellations.
29 The page discusses the potential for propellant delivery in space exploration and commercial satellite fueling, as well as the market for space adventure travel. It

also mentions the potential for space solar power to receive financial and government support.

30 Various organizations and countries, including Air University, CalTech, China, and Japan, are investing in the development of space-based solar power (SSP) technology. This technology has the potential to provide power for in-space needs, military purposes, remote communities, and disaster response. Philanthrocapitalists and strategic investors are the most likely partners to co-invest with the USG in SSP systems. Traditional investors are not likely to invest in first generation reusable launch vehicles (RLVs) due to

31 Private investors, including Google founders and Fidelity, have invested millions in space companies like SpaceX and OneWeb. Philanthrocapitalists and strategic investors are likely to lead the industry, while non-traditional partnerships using Other Transaction Authority (OTA) agreements have proven successful in developing new space capabilities. These partnerships are more cost-effective than traditional methods.

32 The page discusses the potential for cost reduction in the development of ULCATS systems and highlights the DOD's OTA authority for partnerships. It also emphasizes the importance of a portfolio approach to commercial partnerships in lowering program risk. Additionally, it states that international space law allows for military usage of outer space, with certain restrictions.

33 The page discusses the definition of weapons of mass destruction and the prohibition of military bases on celestial bodies. It also highlights the importance of leading in outer space development to establish legal principles and protect national security interests.

34 ULCATS is crucial for national security and economic growth. The military's role in strengthening the industrial base is debated, but having a competitive advantage in space access is essential. Space commerce and power are interconnected, and ULCATS will enable the US to project power globally.

35 Investing in commercial space development can protect national security and create economic benefits. National-level leadership is needed to achieve these goals and leverage soft power for US influence in world events.

36 The page discusses the benefits of ULCATS, including economic growth, improved internet access, environmental monitoring, and space travel. Traditional US government acquisition methods are unlikely to effectively partner with the commercial space industry.

37 The page discusses the challenges of integrating commercial innovation into US government agencies' mission assurance methods for space launch. It recommends partnering with commercial firms, integrating fast space and reusable launch vehicles into the acquisition process, and shaping the interagency environment to maximize the benefits of ultra low cost access to space.

38 Create a purpose-built organization called the NewSpace Development Office (NSDO) to manage commercial ULCATS efforts. This organization should have a fail-fast, fail-forward culture and be located in communities where commercial innovators reside. It should have specific personnel authorities, a sufficient budget, and legal and procurement authorities for rapid RDT&E. A strong leader is needed who embraces these cultural elements and has a proven track record of delivering results.

39 The Air Force is recommended to partner with industry and invest in research and development for next-generation reusable launch vehicles (RLVs) to gain an advantage in the space domain. This will require establishing a purpose-built

40	organization, shaping policy, and championing requirements for RLV systems. Failure to act risks losing America's lead in the space industry.
40	The USAF risks losing national security benefits and market share in critical industries if it does not partner with industry early. However, there is an opportunity for fully reusable access to space that has never been more accessible.
41	The page discusses the need for ULCATS (Ultra Low-Cost Access to Space) in order to make Big LEO (Low Earth Orbit) constellations financially viable for providing affordable internet access. It also highlights the differences between previous failed attempts and current efforts led by private sector entrepreneurs.
42	The page discusses the limitations and lessons learned from previous attempts at developing reusable launch vehicles, such as the Space Shuttle and X-33/VentureStar. It emphasizes the importance of private industry and proper incentives in lowering costs and advancing technology.
43	The page discusses the factors that contributed to the high cost of the National Aerospace Initiative (NAI) and explains why a traditional business case for a reusable launch vehicle (RLV) may not be sufficient. It also refutes the argument that high flight rates would lower launch costs for expendable vehicles as much as for reusables.
44	True reusability of rockets leads to higher reliability and lower operational costs. Industry is willing to partner with the USG for development without early commitments to buy services, but if only one company develops ULCATS systems, prices may rise. OTAs have been successful in developing new rockets in America.
45	The US government should be involved in a private industry decision because it has implications for national security and the government is already heavily involved in space development. This aligns with the trend of transferring responsibility to private industry.
47	The page provides recommendations for proactive approaches to interagency and national policy shaping in areas such as international space law, national leadership, and commercial policy and regulations. It suggests American leadership in establishing space law based on Western values, the creation of an interagency organization overseen by the White House for space-related investments, and transitioning public safety at launch ranges to a civil agency.
48	The page discusses the need for restructuring commercial launch licensing and spacecraft licensing to support the future of space transportation. It also emphasizes the importance of improving collision models and researching debris remediation technologies. The current regulations and policies are not designed for the advancements in space transportation, and changes are necessary for a successful ULCATS marketplace.
49	The legal and regulatory systems for commercial space launch are unable to support the current growth trend in frequency of launch or spacecraft activity, and significant changes are needed to accommodate future demand.
50	Partnerships with US commercial firms pursuing ultra low cost access to space (ULCATS) could enable the Air Force to leverage emerging technologies and impose significant costs on adversaries. ULCATS offers potential for economically affordable and technically feasible new architectures and concepts, as well as prompt global strike capabilities. There is strategic common ground between private space industry developments and USAF needs. The primary barrier to RLV development is the commercial business case.
51	Multiple fully-reusable launch vehicles could lower prices by 3X in the near-term, but achieving a 10X reduction is possible in the long-term. Commercial OTA partnerships are successful and lower cost than traditional methods. A portfolio

	approach to commercial partnerships will lower program risk and increase success likelihood.
52	The page discusses the need to integrate fast space and reusable launch vehicles (RLVs) into Joint requirements and acquisition processes. It also highlights the importance of shaping the interagency environment and national policy, as well as establishing operational legal principles in space. Additionally, it suggests creating a purpose-built organization to effectively partner with the commercial space industry for managing commercial ULCATS efforts.
53	The page discusses the need for new technology in aircraft operations, the prioritization of investment in next-generation technologies, the flexibility of international space law, the importance of American leadership in shaping space law, and the need for national-level leadership in achieving ULCATS.
54	The page discusses the need for regulatory reforms in the commercial space industry to support ULCATS and improve safety at launch ranges. It also highlights key cost drivers and the unlikely success of expendable launch vehicles.
55	The page lists the members of the Air University Assessment Team, including individuals from various organizations and institutions such as NexGen Space LLC, National Defense University, NASA, and Naval Research Laboratory.
57	This page provides a list of acronyms related to air and space operations, including terms such as A2/AD, C2, DARPA, NASA, and RDT&E.
58	This page contains a list of acronyms related to space and the United States Air Force.

Notable Passages

3[2] "Recent private sector developments in access to space could open the door for a new concept for airpower. If realized, these capabilities could fundamentally change the USAF's power projection paradigm, while building new strategic options for the nation."

4 "A Fast Space architecture, marked by rapid reconstitution of proliferated constellations and on-demand user-defined engagement, could leap past the conditions of conventional stalemate built by our competitors."

5 "The ability to immediately deliver additional effects worldwide such as precision navigation and timing, electronic warfare, cyber effects, directed energy, kinetic attack, and rapid global transport of cargo and personnel."

6 "Our analysis has scanned the horizon of innovators across the globe who are experimenting with this concept. The research reveals that reusability is where the big investors are placing their bets. Our RLV analysis finds that launch costs reduce dramatically as launch rates increase. As Figure 1 indicates, as launch rates increase, costs drop quickly and significantly."

7 "Create a purpose-built organization to manage partnerships with commercial ULCATS efforts: To succeed, the USAF needs to create a purpose-built organization, notionally called the 'NewSpace Development Office' (NSDO), which utilizes innovative acquisition processes and methods. This organization requires a 'Fail-Fast, Fail-Forward' culture as opposed to operationally focused cultures where 'failure is not an option.'"

8 "Like the early history of aviation, the coming transformation in affordability of space access has the potential for major breakthroughs in national security and industry that will affect all of humanity."

9 "Groups have become powerfully disruptive actors as well. Leveraging the enmity of the disaffected and the speed of modern information, radical Islamists use religion as a motivation and mayhem as a weapon. The result is globalized, open-ended, information-based conflict that spreads so rapidly and unpredictably that governments have difficulty reacting in time."

10 "In short, competition in space has returned. Space is congested and contested and our advantages in space can no longer be assured. The US military depends on space assets that are increasingly at risk of attack. Further, America's ability to project power globally to defeat and deny aggression is now in question. Our entire long-term defense strategy is being challenged. It is imperative that the US establish first-presence to shape the global policy and establish international precedent that will bring security, predictability and the rule of law to this domain."

11 "A Fast Space architecture specifically addresses the challenges to the three pillars of strategic stability noted by Secretary Work. Nuclear Deterrence. To strengthen our nation's nuclear deterrent, Fast Space facilitates the disaggregation of strategic warning assets from tactical and operational capabilities. The space architecture could be bifurcated into high-end assets for strategic warning, complemented by a resilient rapidly reconstituted constellation of tactical and operational capabilities. This move to disaggregate enables the establishment of clear red lines for our strategic assets. It makes one asset class operationally vulnerable, in policy and in fact, while making strategic satellites the policy

---

[2] Page numbering as supra.

	equivalent of sovereign territory—attacks on which trigger overwhelming and devastating responses."
12	"Advanced space-based capabilities will provide America with new, affordable methods to deter global conflicts, defend the US homeland, enable our international partners, build and develop global relationships, and when necessary decisively defeat our enemies. A Fast Space architecture ensures the US can shape the strategic space environment to our enduring advantage."
13	"First, titans of industry are now fully involved in exploring space and pursuing entrepreneurial ventures. Space is no longer the exclusive province of wealthy governments. Blue Origin, SpaceX, Vulcan Aerospace, and Virgin Galactic headline a growing number of private-sector ventures that see space as the next big thing."
14	"An infusion of government investment and commitment could jump start a commercial innovation cycle that leads to higher flight rates, decreasing costs, reducing entry barriers for more companies, further increasing demand and higher flight rates, thus reducing costs further. To make Fast Space a reality by breaking the cost equation, the US government will need to jump-start this virtuous cycle."
15	"A Fast Space architecture does not solve every problem, but it compellingly addresses many areas of existential concern. A sortie-on-demand launch capability, matched with user-defined real-time engagement, could offset adversary investments and provide new options to the Joint Force and national command authorities."
16	"While the national security requirement to support human spaceflight is less urgent, the inherent obligation of the US government to guard and protect American citizens and resources is clear. As a growing number of private companies lead a transformation in space access, a shift is underway to a new world where American commercial industry, and American citizens, will establish permanent presence in the domain of space. The technologies and systems currently under development to support human space-flight will create new opportunities for the DOD and enable compelling new, cost-effective methods of power projection for the US Joint Force, and global partners. For these reasons, it is imperative that the US government and DOD partner with commercial industry to fully realize the benefits available to the warfighter."
17	"With this capability, the USAF could lead in providing every soldier or marine, every tank, ship, and cockpit access to communications pathways, enhancing situational awareness and creating a combat C2/ISR 'cloud'. This would enable decisive effects for the joint team and nation. The flexibility of global broadband enhanced with software-defined radios creates the potential to create ad hoc command and control structures to better integrate allies and partners."
18	"In the far term (10+ years), new entrants in the launch industry are focused on putting large numbers of humans into space and creating an in-space economy. New supporting space services such as propellant resupply, extraterrestrial resource extraction, on-orbit construction and assembly, and satellite servicing are now attracting significant private investment. This could lead to new national security capabilities including very large apertures with large amounts of available power, the ability to rapidly maneuver in orbit without regret, and the ability to rapidly upgrade and repair satellites in orbit. We should shape our planning and investments to prepare for these coming innovations, and for the eventuality of large numbers of American private citizens living and working in space."

19. "Fast Space will provide the Air Force with the 'the ability to operate in and across air, space, and cyberspace to achieve varying levels of domain superiority over adversaries seeking to exploit all means to disrupt friendly operations' within 45 minutes anywhere on Earth."

20. FINDING [F.1.3]: ULCATS could provide a Stabilizing Deterrent to War: Low-cost responsive space access systems will have the capability to rapidly reconstitute pre-manufactured and stored satellites in Earth orbit. The sheer existence of these ULCATS systems and the ability to rapidly reconstitute our satellites could eliminate an adversary's incentive for preemptive attack. This effect could mitigate the risk of a "Pearl Harbor in space" and create a stabilizing deterrent to war. Weakness invites aggression. America's dependence on space is well known by our potential adversaries. General John Hyten has commented4 "right now we have a very small number of satellites on orbit and our adversaries know exactly

21. "Although this report is not the right place to define specific requirements, our analysis suggests the maximum benefit to the warfighter would come from launch vehicles with the following characteristics. Fully reusable, Two-Stage-To-Orbit. Aircraft-like operability with rapid turn-around between flights. Vertical Takeoff and Vertical Landing (VTVL) supporting small footprint operations to minimize dependence on fixed and vulnerable runways. Standardized first and second stage interfaces. Standardized interfaces for strike, C2, ISR, mobility and spacelift payloads. Operationally significant payload capacity to orbit (e.g., from DARPA XS-1 up to Evolved Expendable Launch Vehicle (EELV) class). This full capability can potentially be demonstrated in as little

22. "The capabilities provided by RLVs do not neatly align in any one Core Function Lead Integrator (CFLI). Private firms are developing game-changing technology that could alter the way in which the Air Force achieves global vigilance, power, and reach. Fast Space and ULCATS enables compelling military benefits across a spectrum that includes C2, ISR, CPGS, Ballistic Missile Defense, PNT augmentation, and SSA. However, no part of the USAF is responsible for all those capabilities. However, there is no single champion to bring these capabilities together in the strategy, planning, and programming."

23. FINDING [F.1.7]: The primary barrier to RLV development is the commercial business case: The primary barrier to 100% private development of an RLV is sufficient proven market-based demand to justify the large high-risk private investment.

24. "Lowering launch costs by 10X to achieve ULCATS requires jump-starting a virtuous cycle (see Figure 2) of industry competition with commercial RLVs that open up new markets. Initially, costs can be reduced by 3X. As new markets and applications develop based on the availability of 3X lower launch costs, this will increase flight rates, and the availability and reliability of launch services. This will increase investor confidence, driving investments in the next generation of RLVs, with increased reliability, robustness, and operability, lowering costs and increasing flight rates even further. This cycle could achieve a 10X reduction in launch costs."

25. "An effective ULCATS strategy will address cost drivers from all of the following:
    - Barriers to driving up flight rate in new markets and in financing new systems
    - Insuring sufficient levels of industry competition
    - Policy, legal, and regulatory barriers
    - Barriers to integrating current technology (partnership tools and methods), and developing future technology for the next generation of systems"

26 "The benefits derived from reusability and aircraft-like operations will be passed on to end users as lower costs and/or higher quality services. Lower costs and higher quality services should drive demand for more space segment capacity and thus higher flight rates. These higher flight rates combined with reusability allow lessons learned to be engineered back into subsequent vehicle models. Better vehicle designs can generate additional operational cost savings through lower labor intensity for maintenance and refurbishments. The virtuous cycle starts all over again, as the process attracts increasing levels of private capital investment."

27 "In the Long-Term, a successful virtuous cycle can enable a launch price reduction of 10X: There is no fundamental physical or economic barrier to an order of magnitude reduction in launch costs—and even more if true aircraft-like operability is achieved. With a virtuous cycle of new markets, higher flight rates, more investment, spurring ever more advanced technologies, leading to reductions in the labor intensity of operations, a 10X reduction or more is reasonably achievable. In the long-term, the primary floor on launch costs could become the cost of propellant, or the non-propellant source of energy used for launch."

28 "Our business case analysis shows that the combination of low prices paid by global broadband customers, the lack of customers over 90% of the planet, and the continuing high-cost of launching satellites, makes it unlikely that traditional investors will invest the billions of dollars needed to finance these systems without significant launch cost reductions."

29 "Of the commercial use cases we evaluated, this one is the furthest along in attracting private capital and developing RLV systems. Of the near-term markets, this has the greatest potential for driving higher flight rates and transforming how we think about and use space."

30 "A successful demonstration could lead to pilot systems of 10-150 Megawatts that might provide services for in-space power, price-insensitive military needs, remote communities, and emergency disaster response."

31 "Philanthrocapitalists and strategic investors will lead: The primary candidates to serve as partners with the USG are the major industry participants, including aerospace primes and several newer companies founded and supported by space enthusiast billionaires. These industry players have the deep pockets, longer investment horizons, technical competence, human resources, and strategic visions to act as effective lead partners."

32 "A portfolio approach to commercial partnerships, on both demand and supply sides, will lower overall strategic program risk and has the highest likelihood of success: The results of a detailed analysis in this study indicate that a commercial development strategy — modeled after the lessons learned from NASA's COTS program, and the DOD's commercial OTAs to develop the EELVs — is well positioned as the catalyst needed to provide transformational breakthroughs in significantly lowering the cost of access to space. The cost reductions available from commercial partnership methods make it possible to afford a portfolio of partnerships."

33 "These examples demonstrate that the 'first movers' in space development will be in a position to impact the development of international space law through their activities. The United States will be in a position to extend its notion of property rights, individual liberty and freedom, and other core concepts of Western civilization and culture, only if the United States takes the lead in developing outer space."

34	"In peace, [strategy]…may gain its most decisive victories by occupying…excellent positions which would perhaps hardly be got by war." --Alfred Thayer Mahan
35	"Space power is a national power multiplier, not through military competition, but through the soft political leverage attained through commercial development."
36	"We can expect that asymmetric warfare will be the mainstay of the contemporary battlefield for some time…But these new threats also require our government to operate as a whole differently -- to act with unity, agility, and creativity. And they will require considerably more resources devoted to America's non-military instruments of power."
37	"USG agencies that implement these important methods of mission assurance have cultures, processes, and values that are in complete alignment with this philosophy of space launch. These same processes, cultures and values — which are critical to these agencies' ability to eliminate the residual risk from expendable launch vehicles — are showstopping-barriers to the commercial innovation process."
38	"A surprising number of innovations fail not because of some fatal technological flaw or because the market is not ready. They fail because responsibility to build these businesses is given to managers or organizations whose capabilities aren't up to the task. Corporate executives make this mistake because most often the very skills that propel an organization to succeed in sustaining circumstances systematically bungle the best ideas for disruptive growth. An organization's capabilities become its disabilities when disruption is afoot."
39	"The USAF stands at a moment of opportunity. Due to the innovations of our private sector, we are briefly ahead of our strategic competition. As Airmen who seek advantage in the space domain, we must champion this opportunity to build an enduring architecture of advantage. A commanding lead in RLVs or any approach to Ultra Low-Cost Access to Space (ULCATS) constitutes an important counter-move to the A2/AD strategies of our adversaries."
40	"A failure to partner early with industry slows the USAF OODA loop to apply these technologies for C2, ISR, mobility, and power projection even while our adversary is taking these steps."
41	"Unfortunately, the business case for the Big LEO constellations still does not close. Sophisticated investors remember the promise and the bankruptcies of Iridium, Globalstar and ORBCOMM, and the collapse of the Teledesic, Skybridge and ICO ventures. Sophisticated investors know the real reason 3 billion people still don't have Internet access is because these people can't afford to pay enough to justify a commercial investment to serve their needs. This is the reason that no large investor, including Google and Facebook, has financed a LEO constellation broadband system."
42	"America has proven over its history that private industry, properly incentivized in a pro-competition environment, is much more successful at lowering costs. As a result it is private industry that designs our automobiles, airplanes, trains and general transportation systems, while the USG serves more effectively as a customer, regulator, and investor in next generation technologies."
43	"We agree that the pure commercial business case — based on the traditional risk-adjusted return on investment (ROI) — does not close. The early days of aviation (air mail) and railroads (transcontinental railroad) had the same problem. In all these cases, the benefits to national security, and to the public and society were so large, that it justified government action to accelerate private development."
44	"After an airplane completes acceptance testing, its reliability to do the job in its flight regime is validated. Expensive tests, maintenance and inspections are no

	longer required between every airplane flight. Instead, we have automated health maintenance systems that watch many indicators for signs of wear and tear. This allows the operator to fuel up, and rapidly fly again, with minimal labor (operational) costs between flights. The same 'aircraft-like operability' benefits will be gained by RLVs as well, and this is where most of the cost savings will be found that lead to ULCATS."
45	"The implications of ULCATS have tremendous near-term and long-term implications to US national security, which means that it is too risky to leave this solely to the decisions of private markets. Further, the USG is already tremendously involved in all space development projects and activities. In the first 50 years of the Space Age, the USG has completely dominated decisions about investment and operations in space. On the spectrum of USG actions, this proposal is consistent with recent trends in the transition of transferring more responsibility for development and operations to US private industry, which some call privatization."
47	"American leadership is vital to creating international space law based on Western values: The United States needs to lead in utilizing space and space resources, in order to establish common international operating principles based on Western values and law. Ceding the "high ground" to other nations to develop this nascent area of the law would enable precedents that directly conflict with Western beliefs and ethics, which could have devastating long-term repercussions to national security, economic policy, and fundamental American freedoms and liberties."
48	"All paths to a future with ULCATS require significant changes to commercial regulations and policies, as current policy and regulations were designed in a world that assumed expensive and infrequent launch."
49	"These regulatory bottlenecks may be solvable over the very short term by applying more money to the various agency programs, but no amount of money can add sufficient capacity to these systems to support the volume of activity expected, and the responsiveness required, of ULCATS. Any significant increase in flight rate from the status quo trend (50% or more) will require significant, if not radical, change in the various licensing and operational government organizations."
50	"Partnerships. The capabilities provided by ULCATS/Fast Space could enable the Air Force to leverage emerging commercial technologies and investments, thereby impose significant costs on adversaries across all five of its core missions through operational agility"
52	"Because much of international space law is based on customary international law, whoever leads in space development is likely to establish operational legal principles, values and practices on the space frontier that will last far into the human future."
53	"It is the national security interests of the United States to establish the Western principles of free trade and commerce, including free enterprise development and use of space resources, in international common law."
54	"O.4 Labor Intensity of Operations is the Most Ignored Key Cost Driver."

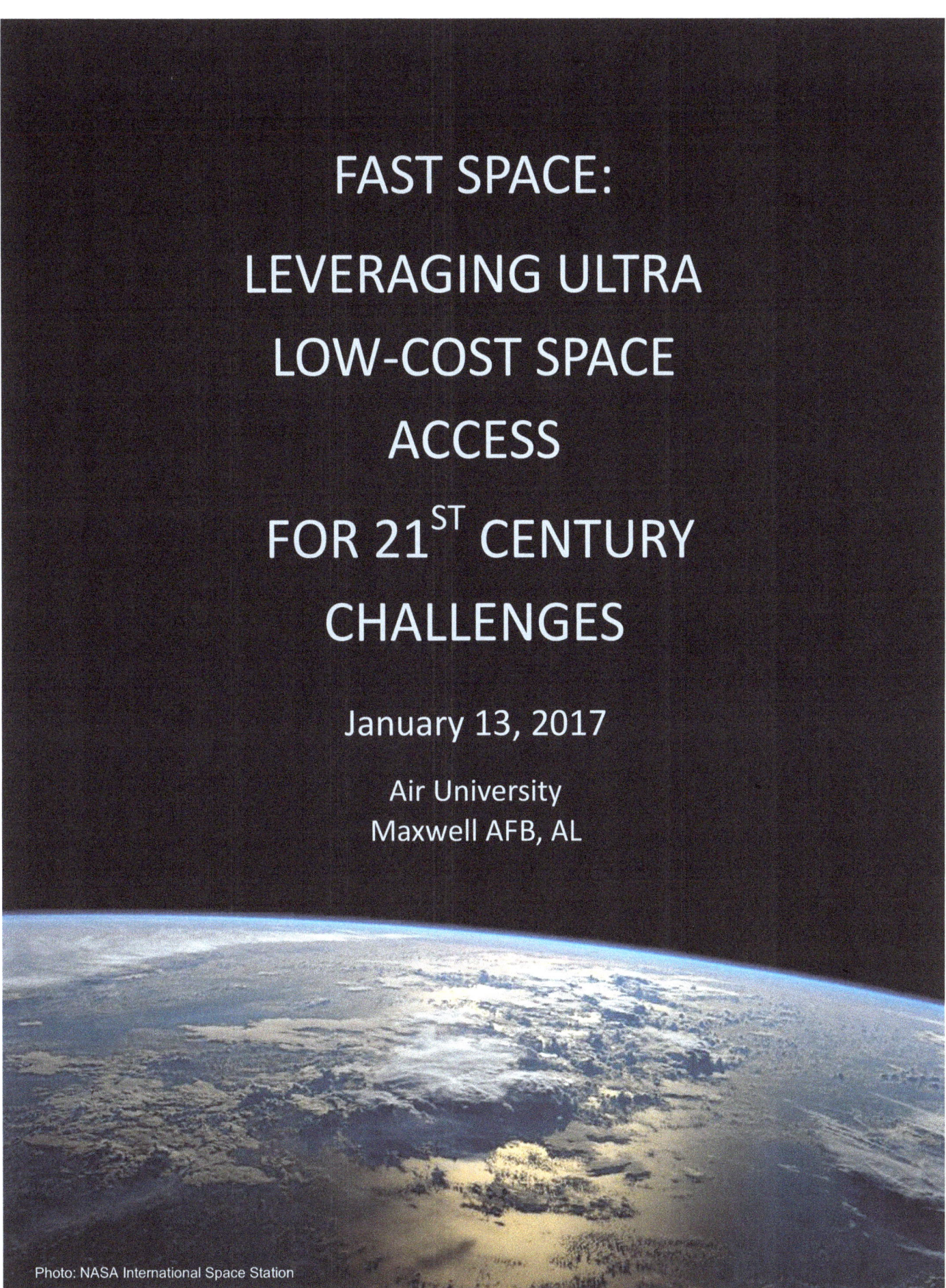

Disclaimer

The views expressed in this academic research paper are those of the authors and do not reflect the official policy or position of the US Government or the Department of Defense. In accordance with Air Force Instruction 51-303, *Intellectual Property, Patents, Patent Related Matters, Trademarks and Copyrights;* it is not copyrighted, but is the property of the US Government.

Air University

Table of Contents

TABLE OF CONTENTS I
FOREWORD II
EXECUTIVE SUMMARY 1
INTRODUCTION 5
    A CHANGING GEO-STRATEGIC LANDSCAPE 5
    RISKS TO INTERNATIONAL STABILITY DUE TO INCREASED COMPETITION IN SPACE 6
    FAST SPACE AS A STRATEGY: A NEW AXIS OF US ADVANTAGE? 7
    STRENGTHENING THE PILLARS OF STRATEGIC STABILITY WITH FAST SPACE 8
    OBSTACLES TO A FAST SPACE APPROACH 9
    CHANGING CONDITIONS YIELD NEW OPPORTUNITIES 9
    REDUCING LAUNCH COSTS 10
    MAKING THE CASE FOR ULCATS TO SUPPORT A FAST SPACE STRATEGY 12
OVERALL FINDINGS 13
    BENEFITS TO THE WARFIGHTER 13
    INDUSTRY VIEWS 20
    TECHNICAL FEASIBILITY ASSESSMENT 20
    LOWERING LAUNCH COSTS 21
    MARKET 25
    PARTNERSHIP STRATEGY 28
    NATIONAL SECURITY STRATEGY & POLICY 29
    NATIONAL LEADERSHIP, ECONOMIC, AND SOFT POWER BENEFITS 31
    PURPOSE-BUILT DEVELOPMENT ORGANIZATION 33
OVERALL RECOMMENDATIONS 34
CONCLUSION 36
APPENDIX A — FREQUENTLY ASKED QUESTIONS A-1
APPENDIX B — PURPOSE-BUILT ORGANIZATION IMPLEMENTATION PLAN B-1
APPENDIX C — RECOMMENDATIONS FOR A PROACTIVE APPROACH TO INTERAGENCY AND NATIONAL POLICY SHAPING C-2
    NATIONAL SECURITY STRATEGY & POLICY C-2
    NATIONAL LEADERSHIP C-2
    COMMERCIAL POLICY & REGULATIONS C-2
    COMMERCIAL REGULATION & POLICY C-3
APPENDIX D — FINDINGS, RECOMMENDATIONS AND OBSERVATIONS TABLE D-1
APPENDIX E — ASSESSMENT TEAM E-1
APPENDIX F — ACRONYMS F-1

FOREWORD

The world is being transformed by shifts in regional power balances, more assertive states, the rise of transnational groups, and proliferating technology. The United States Air Force (USAF), as the vanguard of global vigilance, reach and power for America, must therefore constantly question whether its capabilities, posture, and ability to project are aligned and balanced to contend with these realities.

Growing threats to forward bases, today's space architecture, and our capabilities to hold targets at risk, present the Air Force with stark choices. We can double down on a forward based model of power projection or develop a different way to project power that offsets these threats and uplifts the capability of today's force.

Recent private sector developments in access to space could open the door for a new concept for airpower. If realized these capabilities could fundamentally change the USAF's power projection paradigm, while building new strategic options for the nation.

This study, conducted by a team of leaders in industry, research and development, finance, policy and strategy, explores whether and how the USAF can form private sector partnerships to create a virtuous cycle of launch cost reductions of between 3 and 10 times lower than today's costs. Doing so could enable completely new approaches for the Air Force to defend American values, protect American interests, and enhance opportunities to exploit the unique global advantages of the ultimate high ground.

This study looked at the next steps beyond where industry is today and DARPA's XS-1 program. The team was challenged to keep an open mind and explore all approaches that could dramatically reduce the cost of access to space. While we heard about many game changing technologies that have the potential to provide ultra-low-cost access, including scramjets, tethers, beamed propulsion, and gas guns, we found that US industry is making the most significant private investments in fully-reusable launch vehicles using chemical propulsion. A fundamental element of a commercial partnership strategy is to require private industry to co-invest significant private capital, and then let industry lead the system design and choose the technologies they think are ready.

Simply having technology first does not ensure an enduring lead. While the United States was first to develop the airplane, only a year after the Wright Brothers demonstrated flight in Paris, the French ran away with the military application to such a degree that American Airmen went to war in British and French aircraft. America had to spend great blood and treasure to achieve the high ground and kick off the aviation revolution of the 20th century. In a world of fast moving technological innovation, Airmen would be wise to remember this historical footnote.

STEVEN L. KWAST, Lt Gen, USAF
Commander, Air University

EXECUTIVE SUMMARY

Introduction: A Window of Opportunity in the Space Competition

This white paper describes how the US Air Force (USAF) can leverage new concepts and technologies to build compelling strategic advantage through the innovative exploitation of space. A confluence of government research and private sector innovation has opened a window of opportunity for the United States to shift its approach to space—how it is both viewed and used. Capitalizing on this window requires addressing opportunities across an ecosystem of launch vehicles, payloads, spacecraft, industrial base, and the policy-driven regulatory environment.

The Problem: Competition in Space Undercuts Strategic Stability

The United States (US) is dependent on space for power projection, yet our current space architecture grows increasingly vulnerable. Other nations are developing methods to use space in ways that increase this vulnerability. In a 2016 speech to the Air Force Association, Deputy Secretary of Defense Robert Work noted that enduring strategic stability rests on three pillars: strategic deterrence, conventional deterrence, and managing the strategic environment. For the United States, all three of these pillars rely on a robust space architecture—and all are under threat. Additionally, conventional deterrence is becoming more unaffordable and more vulnerable to an adversary who has been studying us and these approaches to power projection since World War II.

Bolstering Strategic Stability via a Fast Space Strategy

If the three pillars of strategic stability are at risk, where can the United States regain advantage? A new way of thinking about space could shift the competition into a new dimension. This study recommends a new approach called *Fast Space*. This concept is an answer to the demand signal of combatant commanders for solutions to intractable global multi-domain problems such as C2, ISR, ballistic missile defense, and many others.

A Fast Space architecture envisions an ecosystem of concepts, capabilities, and industrial partnerships that make *speed* the defining attribute of advantage in space. In this approach, speed describes both the supply and demand sides of the space market. On the supply side, Fast Space envisions sortie-on-demand launch capability, made possible through economically viable business cases, high launch rates, sustainably lower costs, rapid turn-around, and higher reliability from emerging approaches that industry is experimenting with. On the demand side, Fast Space enables users at all levels of conflict, from tactical to strategic, to harvest new advantages in and through space. These advantages include persistent command and control (C2), ubiquitous communications, on-demand Intelligence, Surveillance, and Reconnaissance (ISR), and new axes for kinetic effects.

A Fast Space architecture, marked by rapid reconstitution of proliferated constellations and on-demand user-defined engagement, could leap past the conditions of conventional stalemate built by our competitors. Imagine the following:

- Aviation-like sortie access to space that would allow a President to defend the United States and coalition interests, signal commitment, and establish assured overwatch in hours.

- Affordable payloads and a flexible mindset toward space that is less risk-averse.
- Joint payloads tailored to maximize immediate effect instead of mission duration.
- A robust and resilient communications system that capitalizes on standardized system architectures and data formats, connecting the ISR and effects grids to deliver a multi-domain and multi-purpose C2 system for US joint force and allies.
- A disaggregated network of interoperable ISR systems that augment national capabilities to provide near real-time global effects to joint and coalition force commanders at the strategic, operational and tactical levels. This network will support intelligence preparation of the battlefield, tactical operations, and battle damage assessment.
- The ability to immediately deliver additional effects worldwide such as precision navigation and timing, electronic warfare, cyber effects, directed energy, kinetic attack, and rapid global transport of cargo and personnel.
- An ability to empower allies and partners via a set of tailored applications.
- A rapidly deployable launch-on-demand system that requires little ground support equipment.

Why Fast Space is Different than Similar Historical Promises

Why is a Fast Space architecture an attractive possibility today, when attempts to build it in the past have been considered too costly or too cavalier? Several fundamental conditions are changing simultaneously: (1) significant private sector investment; (2) maturing capability of emerging technological approaches, such as reusable launch vehicles (RLVs); (3) the US government's expanding use of Other Transaction Authorities (OTAs) to break cost equations; and (4) advances in modern manufacturing and engineering collaboration systems.

Predicate for Success: Reduce Costs through Reusability and Increased Flight Rates

The rich benefits of a Fast Space architecture will only be realized if the cost of space launch can be substantially reduced. Our analysis has scanned the horizon of innovators across the globe who are experimenting with this concept. The research

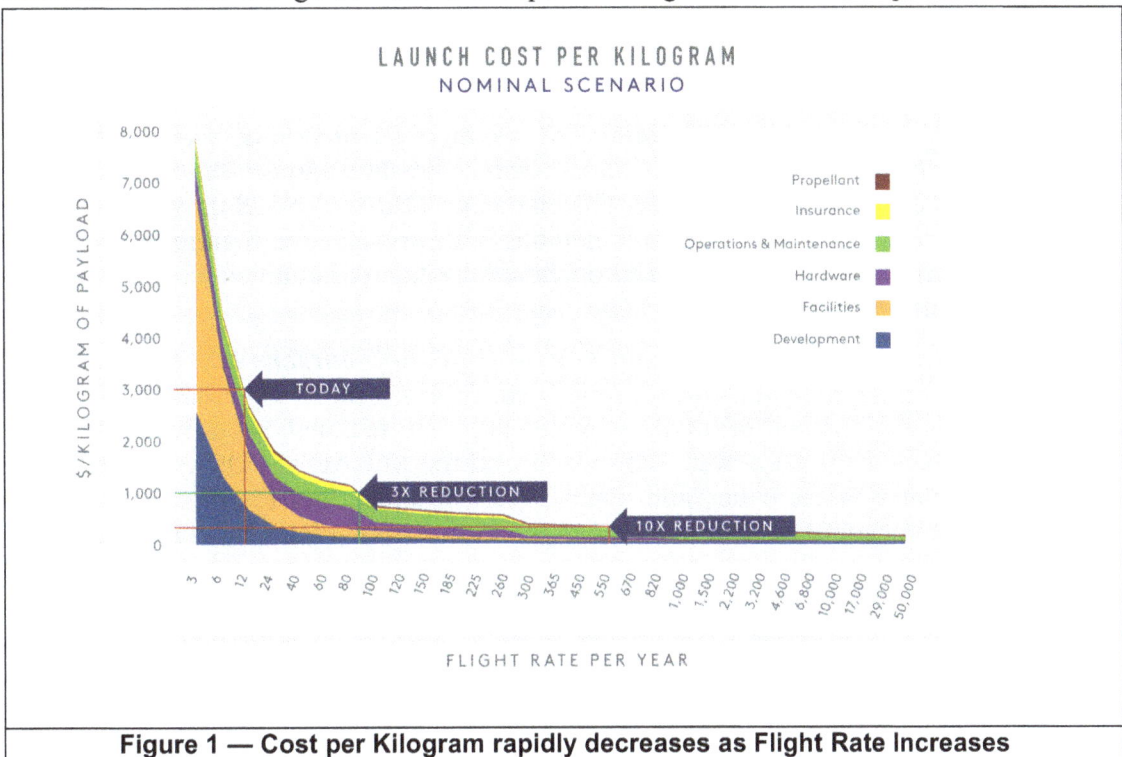

Figure 1 — Cost per Kilogram rapidly decreases as Flight Rate Increases

reveals that reusability is where the big investors are placing their bets. Our RLV analysis finds that launch costs reduce dramatically as launch rates increase. As **Figure 1** indicates, as launch rates increase, costs drop quickly and significantly. The robust analysis that substantiates the figure can be found in the full report. The benefits of Fast Space become real as ultra low-cost access to space (ULCATS) systems mature. Even though RLVs are the current trend, this study recommends that the Air Force ride the leading edge of innovation, no matter where it goes. We cannot predict future winners in this journey, so other technologies such as space elevators, air launch space access, or other techniques may become more affordable.

What the USAF Should Do To Seize This Opportunity

Based on our analysis, we recommend the Air Force should use Other Transaction Authorities (OTAs) to fund commercial partnerships with private space industry leaders. A compelling partnership marries the comparative advantages of both the US government and private industry. The government supplies capital, deep technical expertise and fixed infrastructure beyond the ability of any company to sustain, and the possibility of future purchases if they succeed. Industry capitalizes on their entrepreneurial business models, profit motives, innovative cultures, and extensive research and development to build the technical systems of a Fast Space architecture. A partnership funded through OTAs could put a virtuous cycle of cost reduction into motion to make Fast Space a reality for the joint force.

Key Recommendations

1. ***Partner with US commercial firms pursuing Ultra-Low-Cost Access To Space (ULCATS) using the DOD's Other Transaction Authority (OTA):*** The USAF should assemble a team to pursue the *authority to proceed* with a competition for jointly-funded (cost-shared) prototype OTAs. The full and open competition will seek multiple US commercial partners to develop and demonstrate their proposed space systems in collaboration with USAF financial assistance and broader USG technical resources.

2. ***Create a purpose-built organization to manage partnerships with commercial ULCATS efforts:*** To succeed, the USAF needs to create a purpose-built organization, notionally called the "NewSpace Development Office" (NSDO), which utilizes innovative acquisition processes and methods. This organization requires a "Fail-Fast, Fail-Forward" culture as opposed to operationally focused cultures where "failure is not an option."

3. ***Shape the interagency environment to ease regulatory burdens and lower barriers to entry.*** As the Principal DOD Space Advisor (PDSA), the SECAF has a broad view of how the alignment of civil, commercial, and national security can benefit comprehensive national power. We recommend the SECAF as PDSA take an active stance in maturing the policy and regulatory environment outside the DOD that can maximize the benefit of high launch rate, rapid-turnaround RLVs and associated on-orbit capabilities.

4. ***Integrate consideration of high launch rate rapid-turnaround approaches into the Joint requirements and acquisition process.*** The current process of requirements and acquisition does not incentivize building ground-breaking capabilities. We recommend that relevant DOD organizations create initial capability documents (ICDs) that capture the full suite of opportunities provided associated on-orbit capabilities and champion these to the Joint Requirements Oversight Council (JROC).

INTRODUCTION

On December 17, 1903, Orville Wright piloted the first powered aircraft 20 feet above a wind-swept beach in North Carolina. With little notice or fanfare, the world changed. Nearly 112 years later, on November 23, 2015, Blue Origin's New Shepard launch vehicle made history once again. For the first time a rocket passed the boundary into space and made a successful vertical landing to be reused again. Over the next several months, pioneering space companies traded jabs with a series of dramatic successes. These included the launch and recovery of the reused New Shepard booster and spacecraft and a successful launch and recovery of Space-X's Falcon-9 first stage. Like the early history of aviation, the coming transformation in affordability of space access has the potential for major breakthroughs in national security and industry that will affect all of humanity.

This white paper describes how the US Air Force can leverage new concepts and technologies to build enduring strategic advantage through the innovative exploitation of space. A confluence of government research and private sector innovation has opened a window of opportunity for the United States to shift its approach to space.

Capitalizing on this window requires addressing opportunities across an ecosystem of launch vehicles, payloads, spacecraft, industrial base, and the policy-driven regulatory environment. It provides an opening to uplift American industry, expand the capabilities of coalition partners, revolutionize power projection capabilities of the Joint Force, and enhance our national security. However, the window to seize the initiative is limited. Other space-faring competitors are moving quickly to duplicate this technology.

This study, conducted by a team of leaders in industry, science, finance, policy, and strategy, proposes a "Fast Space" concept that envisions a new architecture of asymmetric capabilities, fueled by sortie-on-demand reusable launch, enabling rapid user-defined applications for the Joint Force and coalition partners. The paper defines a series of steps the USAF can take today to make these concepts reality and details a list of associated recommendations. These recommendations include a decision to partner with industry, employing a purpose-built organization, seeking to proactively shape the interagency policy environment, and actively addressing the requirements-driven acquisition process.

A Changing Geo-Strategic Landscape

The post-Cold War geo-strategic environment is changing. A world order that was once remarkable for its degree of global cooperation and adherence to norms now sees growing competition and mistrust. New forms of competition and conflict involving both states and groups blur the distinction between peace and war.

While globalization remains the engine of economic prosperity worldwide, a growing number of actors seek to carve spheres of influence within the globalized system by gaining positional, legal, and informational advantage.[1] Meanwhile, these actors seek to

---

[1] For more information on China's regional strategy see Michael Pillsbury, *The One Hundred Year Marathon: China's Secret Strategy to Replace America as a Global Superpower* (New York: St Martin's, 2015). For information on Russia's hybrid war strategy in its near abroad, see John R. Haines, "How, Why, and When Russia Will Deploy Little Green Men – and Why the US Cannot" Foreign Policy

check the US military's ability to project power into their sphere. They employ asymmetric counter-intervention strategies designed to increase US force requirements while denying forward basing and use of space capabilities. This threatens to undercut United States assurance to partners and allies.

Groups have become powerfully disruptive actors as well. Leveraging the enmity of the disaffected and the speed of modern information, radical Islamists use religion as a motivation and mayhem as a weapon. The result is globalized, open-ended, information-based conflict that spreads so rapidly and unpredictably that governments have difficulty reacting in time.[2]

Risks to International Stability Due to Increased Competition in Space

Given this new reality, today's leaders face high-stakes decisions on how to enable US power projection and coalition partnership. Today, the United States (US) is dependent on space for power projection, yet our current space architecture grows increasingly vulnerable. In a 2016 speech to the Air Force Association, Deputy Secretary of Defense Robert Work noted that enduring strategic stability rests on three pillars: strategic deterrence, conventional deterrence, and managing the strategic environment. For the United States, all three of these pillars rely on a robust space architecture—and all are under threat.

Insufficient Leverage from Nuclear Deterrent. First, the logic of nuclear deterrence applies unevenly across the international system. Many states or groups feel few if any constraining effects from the US nuclear deterrent. Further, US satellites that provide strategic warning of nuclear launch are increasingly at risk of attack, undermining the awareness that nuclear stability requires.

Conventional Advantage Checked by Adversary Investments. Second, our military operations over the past 25 years have been closely studied by powerful states. China and Russia, for example, have invested heavily in a portfolio of military capabilities that blunt the razor edge of US military power. Our infrastructure of forward bases and exquisite satellites is increasingly at risk of devastating attack, and our ability to hold targets at risk is limited in key regions of the world. Even in permissive environments, the US military's operational approach relies heavily on space for command and control, communications, and intelligence. Our current approach to conventional power projection is vulnerable, economically unsustainable, and rendered ineffective in regions of great interest.

Changing Dynamics in Space Environment. Third, the strategic environment in space is changing rapidly. Space is no longer dominated by the US government, as both private industry and other strong states accelerate their efforts to explore and exploit the domain. Billionaire philanthrocapitalists like Elon Musk, Jeff Bezos, Paul Allen, and Richard Branson are venturing personal fortunes in private space enterprises. Men who have earned billions of dollars seizing timely opportunities have all turned their attention

---

Research Institute, March 9, 2016, http://www.fpri.org/article/2016/03/how-why-and-when-russia-will-deploy-little-green-men-and-why-the-us-cannot/.

[2] For more on the rapid rise and proliferation of radical Islamist thought and its challenge to western governments, see William McCants, *The ISIS Apocalypse: The History, Strategy and Doomsday Vision of the Islamist State* (New York: St. Martin's, 2015).

to space—the alignment of their collective vision suggests profound opportunity. Their collective efforts will soon launch US citizens into space as tourists and entrepreneurs. When they do, will US military protection extend to our citizens in space? When other states build an enduring presence in space—which they are actively pursuing—will the US be content with the standards and norms others establish?

In short, competition in space has returned. Space is congested and contested and our advantages in space can no longer be assured. The US military depends on space assets that are increasingly at risk of attack. Further, America's ability to project power globally to defeat and deny aggression is now in question. Our entire long-term defense strategy is being challenged. It is imperative that the US establish first-presence to shape the global policy and establish international precedent that will bring security, predictability and the rule of law to this domain.

Fast Space as a Strategy: A New Axis of US Advantage?

If the three pillars of strategic stability are at risk, where can the United States regain advantage? A new way of thinking about space could shift the competition into a new dimension.

This study proposes a *Fast Space* approach—an ecosystem of concepts, capabilities, and industrial partnerships that make *speed* the defining attribute of advantage in space. In this approach, speed describes both the supply and demand sides of the space market. On the supply side, Fast Space envisions sortie-on-demand launch capability, made possible through economically viable business cases, high launch rates, sustainably lower costs, rapid turn-around, and higher reliability from RLVs and other emerging approaches. On the demand side, Fast Space enables users at all levels of conflict, from tactical to strategic, to harvest new advantages in and through space. These advantages include persistent command and control (C2), ubiquitous communications, on-demand ISR, and new axes for kinetic effects.

A Fast Space architecture, marked by rapid reconstitution of proliferated constellations and on-demand user-defined engagement, could leap past the conditions of conventional stalemate built by our competitors. Imagine the following:

- Aviation-like sortie access to space that would allow a President to defend the United States and coalition interests, signal commitment, and establish assured overwatch in hours rather than days.

- Affordable payloads and a flexible mindset toward space that is far less risk-averse than today's expensive exquisite approach.

- Commercial development launched to maximize immediate effect rather than mission duration or platform longevity.

- A robust and resilient communications system that capitalizes on standardized system architectures and data formats, connecting the ISR and effects grids to deliver a multi-domain and multi-purpose C2 system for US joint force and allies.

- A disaggregated network of interoperable ISR systems that augment national capabilities to provide near real-time global effects to joint and coalition force commanders at the strategic, operational and tactical levels. This network will support intelligence preparation of the battlefield, tactical operations, and battle damage assessment.

- The ability to immediately deliver additional effects worldwide such as precision navigation and timing, electronic warfare, cyber effects, directed energy, kinetic attack, and rapid global transport of cargo and personnel.

- The ability to deliver worldwide applications like command and control, data relay, precision navigation and timing, electronic warfare, cyber, directed energy, or kinetic effects against area-denial capabilities threatening US forces.

- An ability to empower allies and partners by providing access to a set of tailored applications.

- A rapidly deployable launch-on-demand system that requires little ground support equipment and allows for launch from any airfield into any inclination, complicating space situational awareness for others.

- Resilient communications to preserve battle networks, command and control, and national transactions.

Strengthening the Pillars of Strategic Stability with Fast Space

A Fast Space architecture specifically addresses the challenges to the three pillars of strategic stability noted by Secretary Work.

Nuclear Deterrence. To strengthen our nation's nuclear deterrent, Fast Space facilitates the disaggregation of strategic warning assets from tactical and operational capabilities. The space architecture could be bifurcated into high-end assets for strategic warning, complemented by a resilient rapidly reconstituted constellation of tactical and operational capabilities. This move to disaggregate enables the establishment of clear red lines for our strategic assets. It makes one asset class operationally vulnerable, in policy and in fact, while making strategic satellites the policy equivalent of sovereign territory—attacks on which trigger overwhelming and devastating responses. Furthermore, Fast Space assets could create conventional coercive leverage over groups or states who consider themselves immune to nuclear threats.

Conventional Capability and Deterrent. A Fast Space approach significantly improves our conventional capability, thereby strengthening the deterrent value of our conventional military power. Currently, adversaries have checked our conventional advantage by building anti-access and area-denial (A2/AD) capabilities that make fixed targets like forward bases vulnerable to attack. These A2/AD capabilities severely weaken the persistence, tempo, and lethality that underwrite our current approach to power projection. Current conditions will not permit the US military to maintain a sufficient density of sensors and effects over the required distances and lengths of time.

How could the Fast Space architecture change the game? Fast Space capabilities, integrated with cyber, undersea platforms, and stealthy airborne capabilities, provide a new paradigm for power projection with far less reliance on forward bases. The integration of presence with speed, multi-axis approaches, and multi-domain operations, all synchronized and directed by a retooled command and control construct, could offer an opportunity for continuing advantage.

Fast Space accepts vulnerability as the new global condition and builds resiliency in response. The vulnerability of our current high-end expensive satellites compels the need to spread our risk over a broader portfolio of assets. Vulnerability calls for proliferation of smaller, cheaper satellites that can be replenished almost on demand. Fast Space pursues the ecosystem of partnerships, concepts, and capabilities to make that possible.

Shaping the Strategic Environment. Advanced space-based capabilities will provide America with new, affordable methods to deter global conflicts, defend the US homeland, enable our international partners, build and develop global relationships, and when necessary decisively defeat our enemies. A Fast Space architecture ensures the US can shape the strategic space environment to our enduring advantage. A space launch paradigm with high sortie rates, high reliability, and low costs encourages industry participation, entrepreneurial ventures, and new sources of commerce and economic growth. The US government is well positioned to reduce transaction costs, set common standards, and facilitate integration of related technologies. Furthermore, government investment can strengthen the industrial base, which underwrites national prosperity and our enduring ability to defend our interests. Lastly, a sortie-on-demand capability equips the USAF to protect US citizens anywhere on the globe—and eventually, beyond it. As tourists and industrialists spend more and more time in space, a call for military protection will increase—the Air Force must be ready.

Obstacles to a Fast Space Approach

Fast Space is not a panacea, but it holds tremendous promise. Furthermore, this line of thinking is not new—we have been down this road before. If the strategic case is so compelling, why has this vision not been achieved in the past?

From the Dyna-Soar/X-20 program in the 1960s, to the National Aerospace Plane (NASP) program in the 1980s, to the Military Spaceplane (MSP) program in the 1990s, the Air Force has pursued responsive, ultra-low-cost access to space on several occasions in its history. Despite glimpses of strategic and technical promise, these proposals ultimately failed to achieve the feasibility, momentum, and approval to test the capability.

Several key hurdles have prevented past proposals from achieving fruition: (1) the perceived lack of a compelling military need; (2) high costs of development and acquisition; (3) technical infeasibility of RLVs; and (4) a requirements-driven acquisition process that was not structured for paradigm-breaking capabilities.

When these earlier proposals were considered, the US had sufficient nuclear and conventional superiority; the need for highly-responsive low-cost space access appeared limited. Moreover, even if decision-makers wanted to get to space quickly and reliably, the cost curve did not support the desire. The technology to build RLVs was not sufficiently mature to increase flight rates and drive down the astronomical costs of space launch. Finally, the acquisitions process—from requirements through contracting through systems integration—did not encourage the development of entirely new paradigms.

In sum, getting to space is exceptionally difficult and thus exceedingly expensive. Gravity's burdensome tax—expensive, infrequent, unresponsive and relatively unreliable access to orbit—has created a paradigm through which all applications that use the space environment have been viewed to date.

Changing Conditions Yield New Opportunities

Why is a Fast Space architecture an attractive possibility today, when attempts to build it in the past have been considered too costly or too cavalier? Several fundamental conditions are changing simultaneously: (1) Significant private sector investment; (2)

maturing technologies and subsystems used in RLVs and other emerging approaches; (3) the USG's expanding use of Other Transaction Authorities (OTAs) to break cost equations; and (4) advances in modern manufacturing and engineering collaboration systems.

First, titans of industry are now fully involved in exploring space and pursuing entrepreneurial ventures. Space is no longer the exclusive province of wealthy governments. Blue Origin, SpaceX, Vulcan Aerospace, and Virgin Galactic headline a growing number of private-sector ventures that see space as the next big thing. Along with traditional aerospace firms like Boeing, Northrop Grumman, and Orbital-ATK, companies are competing to see who can build the business model that integrates cost-effective launch, compelling payloads, and value-added customer experiences. As private investment expands, new markets emerge, technologies mature, and costs come down.

A significant cost-flip is already underway in the commercial satellite market. Traditionally, satellite programs were so expensive that launch costs comprised a relatively small fraction of the total program cost. Now, the cost of building small satellites for Low Earth Orbit (LEO) is plummeting. Launch costs now represent a much higher percentage of the business expense, incentivizing efforts to reduce the cost of launch in any way possible.

The second changing condition relates closely to the first. With such widespread private investment, game-changing technologies are beginning to mature. RLVs with rapid turn-around and affordable launch-on-demand are a core requirement of an affordable Fast Space ecosystem, and both Blue Origin and SpaceX have successfully demonstrated early capability in this area. Maturation is still needed, of course, but successful demonstrations are stimulating interest, investment, and new excitement in the once-moribund world of space exploration.

The third promising development is the increased familiarity within the DOD in using OTAs. The OTA vehicle has proven effective to build partnerships with industry that break traditional cost curves. It allows the government to structure partnerships with private industry that look more like traditional commercial methods. Successful private companies who dread the burdensome requirements of the traditional Federal Acquisition Regulation (FAR) directly expressed strong interest in doing business with government through OTAs. Similarly, government agencies have used OTAs to align incentives and share risk with industry partners; doing so has enabled major acquisitions at much lower price points. In short, we now have more compelling proof that OTAs with commercial partnerships can break traditional cost equations.

An alignment of enabling factors suggests the time to act is here. Industry is more involved than ever. They have demonstrated game-changing reusable launch technologies. OTAs have proven effective as a vehicle for public-private partnership (PPP) to bring down cost. How then could a partnership built through OTAs succeed in breaking the cost equation, making the Fast Space vision a reality?

Reducing Launch Costs

The rich benefits of a Fast Space architecture will only be realized if the cost of space launch can be substantially reduced. Our RLV analysis (as illustrated in **Figure 1** of the Executive Summary) finds that launch costs reduce dramatically as launch rates increase. The analysis that substantiates the figure can be found in the study full report.

If costs plummet as a function of flight rates, why has the market not moved in this direction already? In the space industry this has been referred to as a "chicken and egg problem". What comes first, ULCATS that enables very large space markets, or the large markets that need ULCATS? This problem has not been solved yet because the required investment to develop commercial RLVs is large and risky. Future inventions and approaches will inevitably change the game and provide new pathways for us to follow. The only way we can capitalize on these developments is by being a close part of the journey.

While private capital markets do fund comparably-sized investments, they will not do so when the existing markets are not large enough and the new markets are not proven. When traditional investors invest billions in drilling for oil, in developing new drugs or a next generation airplane, in constructing a semi-conductor facility or a large skyscraper, they have a high-degree of certainty about customer demand. Investor perception of market risk is informed by lessons of many large space projects that have failed or gone through bankruptcy (X-33/VentureStar, Rocketplane Kistler, Globalstar, Iridium, Skybridge, Teledesic, ICO, Orbcomm, NewSat, etc.)

Analysis indicates latent market demand exists and the cycle can be reversed. An infusion of government investment and commitment could jump start a commercial innovation cycle that leads to higher flight rates, decreasing costs, reducing entry barriers for more companies, further increasing demand and higher flight rates, thus reducing costs further. To make Fast Space a reality by breaking the cost equation, the US government will need to jump-start this virtuous cycle.

Figure 2 — Virtuous Cycle of Reinforcing Growth in Markets, Innovation and Investments

Figure 2 illustrates the components of the virtuous cycle that government investment could stimulate. As the rest of this white paper substantiates, our analysis argues the Air Force should use OTAs to fund commercial partnerships with private space industry leaders. A compelling partnership marries the comparative advantages of both the US government and private industry. The government supplies capital, deep technical expertise and unique infrastructure beyond the ability of any company to sustain, and the possibility of future purchases if they succeed. Industry capitalizes on their entrepreneurial business models, profit motives, innovative cultures, and extensive research and development to build the technical systems of a Fast Space architecture. A partnership funded through OTAs could put the virtuous cycle into motion.

Making the Case for ULCATS to Support a Fast Space Strategy

The strategic context of the United States requires new thinking and game-changing action. Our strategic dilemmas are legion. Our strategic deterrent does not produce leverage in areas where we need it. In regions of key concern, our conventional superiority has been countered or blunted. Our infrastructure is vulnerable and brittle; we lack agility and resilience in our approach to power projection. A Fast Space architecture does not solve every problem, but it compellingly addresses many areas of existential concern. A sortie-on-demand launch capability, matched with user-defined real-time engagement, could offset adversary investments and provide new options to the Joint Force and national command authorities.

The balance of this paper substantiates and expands this argument further. It highlights the enduring benefits of a Fast Space architecture for our coalition partners and the Joint Force, enabled by ultra-low-cost access to space (ULCATS). The study results are expanded in a full report under the same title as this paper.

OVERALL FINDINGS

Benefits to the Warfighter

Studies associated with earlier programs (e.g., Dyna-Soar and MSP) have repeatedly shown substantial benefit to the warfighter if the US achieves a capability like the Fast Space architecture. Over the course of this study, the team uncovered similar benefits in the significant overlap between developments in the commercial space sector and national security needs.

Strategic Common Ground

The most significant investments underway in the commercial sector are squarely focused on three areas; space-based remote sensing, communications, and private human space flight. Notably, the most immediate needs in the national security sector envision new multi-domain and multi-functional C2 and ISR capabilities to enhance current capacity with disaggregated systems that provide resilient and responsive global operational effects. The study team identified a strategic common ground with significant overlaps as illustrated in **Table 1**.

Commercial Application	Military Application
Large LEO constellations for Communications	Global Dynamic C2, Strategic Integration
Large LEO constellations for Remote Sensing	ISR, SSA
Human Spaceflight (Reusable space access systems that provide very low-cost, much higher reliability, higher flight rates, and rapid turn-around)	Rapid Reconstitution, Rapid Global Mobility, Air & Space Superiority, Global Strike
Table 1 — Strategic Common Ground between Commercial and National Security Space	

While the national security requirement to support human spaceflight is less urgent, the inherent obligation of the US government to guard and protect American citizens and resources is clear. As a growing number of private companies lead a transformation in space access, a shift is underway to a new world where American commercial industry, and American citizens, will establish permanent presence in the domain of space. The technologies and systems currently under development to support human space-flight will create new opportunities for the DOD and enable compelling new, cost-effective methods of power projection for the US Joint Force, and global partners. For these reasons, it is imperative that the US government and DOD partner with commercial industry to fully realize the benefits available to the warfighter.

Capabilities

The private sector is moving towards significant reductions in the cost of launch and is on the path toward a proliferated constellation of small satellites for sensing, communications, and command and control (C2) in space. This creates significant potential for contributions to all five core Air Force missions, as well as to the missions of the entire Joint and Combined force in the near, mid and long term.

In the near term (1 to 3 years), industry will lay critical groundwork for the first tranche of LEO constellations with data and sensor services, such as OneWeb, the ~4000 satellite SpaceX internet architecture, and the 1400-3000 satellite Boeing constellation. These purely commercial "LEO" constellations could create the earliest opportunities for the warfighter in the C2/ISR mission area. But, the business cases of these constellations are unlikely to close with current launch prices and market conditions. By working with industry during this period, the Air Force could leverage billions in private sector investments and help industry close their business case. In addition to buying services from commercial constellations, opportunities also to procure cost-effective government satellites, or government payloads, that can be integrated with commercial systems. The result would be affordably networking the effects and ISR grids with a series of global, multi-domain and multi-purpose C2 and ISR constellations.

Potential benefits for DOD during this initial partnership are two-fold: 1) the early USAF role would ensure the common architectures, mission systems, and standardized data formats would be compatible with future DOD systems, and 2) the DOD could harvest initial operational capabilities from the first spacecraft launched in these new constellations. This approach would provide a framework for the USAF to transform global C2 capabilities, enabling decisions to be made more rapidly at the strategic, operational and tactical levels.

With this capability, the USAF could lead in providing every soldier or marine, every tank, ship, and cockpit access to communications pathways, enhancing situational awareness and creating a combat C2/ISR "cloud". This would enable decisive effects for the joint team and nation. The flexibility of global broadband enhanced with software-defined radios creates the potential to create *ad hoc* command and control structures to better integrate allies and partners.

Near term opportunities also exist for applications of the Mach 3 New Shepard and Mach 10 Falcon 9 Reusable (F9R) first stage. These Vertical Takeoff, Vertical Landing (VTVL) suborbital RLVs from Blue Origin and SpaceX have flown multiple demonstration flights to date and could be quickly modified to carry military ISR and other payloads.

In the mid-term (3 to 10 years), the emergence of "Fast Space" in the private sector could provide other significant benefits to USAF and Joint missions. This is the period where the large LEO constellations that leverage commercial RLVs that are at least 3X lower in cost are likely to emerge. In addition to the C2 and ISR benefits, with relatively minor modifications, a "Big LEO" constellation could become the basis for a much more resilient global positioning, navigation and timing (PNT) system. Such a system would both augment and backstop GPS in a contested space environment. With Fast Space, a proliferated PNT constellation would be much more resilient and quickly reconstituted than existing GPS—reducing our vulnerability to jamming and increasing the effectiveness of the entire joint force. It would also reduce an adversary's incentive to attack the existing GPS system preemptively.

In addition, potential benefits such as space-based electronic attack and electronic protection have the potential to enhance the effectiveness and survivability of Joint and Combined forces. This is well aligned with the CSAF-chartered *Air Superiority 2030 Flight Plan,* which notes that by 2030, increasing lethality and reach of adversary weapons will significantly increase the risk to large battle-management platforms such as Airborne Warning and Control Systems (AWACS). These threats will limit current platforms' ability to see and manage activities in contested environments. To overcome these shortfalls, the 2030 Flight Plan directs the Air Force to develop concepts that disaggregate this capability using multiple sensor platforms, including teamed manned and unmanned systems, a robust battlespace information architecture, and dispersed command and control. It also seeks increased contributions from space-based assets so the Air Force and the Joint Force can increasingly rely on advantages provided by on-orbit assets for air superiority. C2 and ISR provided by Fast Space can be a significant part of executing the *Air Superiority 2030 Flight Plan*.

An early investment in Fast Space also places the Air Force on a clear path toward the capabilities described by the *Air Force Future Operating Concept* (AFFOC). The AFFOC establishes the following vision for the force of 2035:

- High-end AF vehicles can transit through the space domain to deliver effects at hypersonic speed across global distances
- AF spacelift capabilities provide timely launch to support mission needs for agile space capabilities
- Execute both surface-launched and air-launched space lift missions to transport materials, assets, and personnel to space
- Mobility operations in space expanded to include transport of personnel and servicing of space assets, be it fuel replenishment or repair/mission enhancement through replacement of mission modules and the movement of space assets within or between orbits.

The AFFOC also seeks space launch capabilities that enable flexible employment of short-term space capabilities, including space control and reconstitution of lost or degraded space capabilities. It foresees a strong relationship with commercial partners through a "Civil Reserve Space Fleet" to supplement steady-state needs, providing scalable capacity for rapid and responsive space lift to support crisis operations in air, space, and cyberspace. It looks to space to provide revolutionized battle management command, control and communication (BMC3) of air, space, and cyberspace operations to enable prompt, effective multi-domain coordination of effects. Fast Space brings all of these AFFOC capabilities within reach, leveraging private sector investment and innovation to help us get there sooner than 2035.

In the far term (10+ years), new entrants in the launch industry are focused on putting large numbers of humans into space and creating an in-space economy. New supporting space services such as propellant resupply, extraterrestrial resource extraction, on-orbit construction and assembly, and satellite servicing are now attracting significant private investment. This could lead to new national security capabilities including very large apertures with large amounts of available power, the ability to rapidly maneuver in orbit without regret, and the ability to rapidly upgrade and repair satellites in orbit. We should shape our planning and investments to prepare for these coming innovations, and for the eventuality of large numbers of American private citizens living and working in space.

As the Fast Space architecture matures over time, it has the ability to directly support each of the AFFOC 2035 core missions:

- **Multi-Domain Command and Control** — A distributed, resilient capability provided by rapidly reconstituted small satellites could create a persistent communications and data infrastructure over the joint force. Fast Space minimizes the vulnerability of current space and land-based C2 assets by ensuring rapid reconstitution at a time and place advantageous to the United States.

- **Adaptive Domain Control** — Fast Space will provide the Air Force with the "the ability to operate in and across air, space, and cyberspace to achieve varying levels of domain superiority over adversaries seeking to exploit all means to disrupt friendly operations" within 45 minutes anywhere on Earth. By manipulating distance and time through Fast Space, the Air Force will be in an advantageous position to deliver its core mission effects across each domain.

- **Global Integrated Intelligence, Surveillance and Reconnaissance (GIISR)** — Fast Space will aid the Air Force in accelerating the decision-making cycle in a challenging hider/finder world environment. The ability to rapidly deploy ISR assets with the capability to integrate through multiple domains will provide operational agility to the GIISR mission.

- **Rapid Global Mobility** — Global reach advantages of the space domain can leverage existing investments by the private sector in orbital and suborbital capabilities to deliver Air Force core mission effects. Fast Space will allow for an autonomous drone package of ISR, C2, and strike capability to rapidly support a threatened embassy in the USAFRICOM AOR, for example. Currently, response plans in USAFRICOM and USEUCOM involve responding to a threat in Africa with C-130's on alert at Ramstein Air Base, Germany. ULCATS minimizes the intelligence needed to identify a threat while also eliminating long logistics lines and insufficient crisis response times. This will give personnel and assets back to the Combatant Commanders while posturing additional capabilities.

- **Global Precision Strike** — As with rapid global mobility, Fast Space will allow low-cost sortie-based repeatable delivery of strategic effects anywhere in the world on a prompt and sustained basis through waves of repeatable, affordable sorties. Forward operating bases can be minimized to the most advantageous level of posture versus presence to ensure the appropriate amount of deterrence is achieved. Fast Space will enable human-machine combat teaming in order to maximize conventional capabilities with emerging global precision strike technologies.

FINDING [F.1.1a]: The capabilities provided by ULCATS/Fast Space could enable the Air Force to leverage emerging commercial technologies and investments, and thereby impose significant asymmetric costs on our adversaries across all five of its core missions through operational agility.

Without a government program or a set of government requirements dedicated to the development of RLVs, US private industry has built and flown two reusable first stages, the Mach 3 Blue Origin New Shepard and the Mach 10 SpaceX Falcon 9FT Upgrade first stage. It is technically feasible and affordable to use these vehicles today, and they are on a path to even lower costs.

Because these vehicles have not been designed with national security needs in mind, it would require some investment to adapt them to military missions. It will cost far less in both time and money to modify existing RLVs than to develop new reusable first stages from scratch.

FINDING [F.1.1b]: Whole new architectures and concepts that will transform USAF dynamic C2 and ISR could become economically affordable and technically feasible with ULCATS. Because of the revolution in electronics driven by Moore's Law combined with the revolution in commercial small satellites, transformational concepts that were previously unaffordable are becoming obvious applications. For example, it will be economically possible to provide global dynamic C2 and ISR, and jam-resistant in-theater communications by purchasing services from commercial LEO satellite constellations. These systems will be "good enough" for many military applications.

FINDING [F.1.1c]: Commercial RLVs with rapid turn-around could provide the United States a prompt global strike capability. National security studies conducted since the Dyna-Soar effort repeatedly demonstrate the value of being able to strike anywhere in the world from the continental United States with conventional weapons. DOD has also studied time-critical suborbital conventional strike — most recently in the 2008 Defense Science Board Report on *Time Critical Conventional Strike from Strategic Standoff*[3].

FINDING [F.1.1d]: An early success may be theater pop-up missions for suborbital RLVs: Recent demonstrations of commercial reusable first-stages hold out the possibility of affordable systems that can deliver useful effects for the theater commander. The Air Force might overcome existing threats and challenges by leveraging commercial reusable vehicles that are already in flight test, as companies develop mature orbital systems.

FINDING [F.1.2]: There is a strategic common ground between rapid developments in private space industry and USAF needs. Private industry is making rapid progress in the development of more affordable satellites that leverage Moore's Law and rapid progress in consumer electronics. Private firms are attempting to develop large LEO constellations for high-bandwidth communications and remote sensing, which have much in common with national security needs.

FINDING [F.1.3]: ULCATS could provide a Stabilizing Deterrent to War: Low-cost responsive space access systems will have the capability to rapidly reconstitute pre-manufactured and stored satellites in Earth orbit. The sheer existence of these ULCATS systems and the ability to rapidly reconstitute our satellites could eliminate an adversary's incentive for preemptive attack. This effect could mitigate the risk of a "Pearl Harbor in space" and create a stabilizing deterrent to war.

Weakness invites aggression. America's dependence on space is well known by our potential adversaries. General John Hyten has commented[4] "right now we have a very small number of satellites on orbit and our adversaries know exactly where they are. If you know exactly where they are, then it's fairly easy to figure out how to deny the capabilities that come off those satellites."

FINDING [F.1.4]: While Air Force needs for space operations that are "aviation like" are aligned with commercial interests, they are not identical. By co-investing with industry through OTA mechanisms we can influence industrial plans to mutual benefit. In the near term, we can adapt current commercial platforms.

Sitting on the sidelines entails significant risks. These risks include: 1) the possibility that industry, without an active USG partner will develop systems that have moderate or

---

[3] Kerber, Ronald and Robert Stein, *Time Critical Conventional Strike from Strategic Standoff*, Report of the Defense Science Board Task Force, March 2009.

[4] http://www.afspc.af.mil/About-Us/Leadership-Speeches/Speeches/Display/Article/731711/afspc-defending-our-edge

minimal value to the warfighter, 2) that industry will take longer to develop these revolutionary systems, and 3) that other countries will play a more active role in their industry and leap-frog the United States. In the early days of aviation, Europe captured world leadership in flight because they partnered with their industry, while the United States provided relatively minor support to accelerate the work of the Wright brothers and Glen Curtis.

In the longer term, we will need to acquire purpose-built vehicles that more closely meet USAF needs in the same way the Boeing 707 was equivalent to the KC-135. Although this report is not the right place to define specific requirements, our analysis suggests the maximum benefit to the warfighter would come from launch vehicles with the following characteristics.

- Fully reusable, Two-Stage-To-Orbit
- Aircraft-like operability with rapid turn-around between flights
- Vertical Takeoff and Vertical Landing (VTVL) supporting small footprint operations to minimize dependence on fixed and vulnerable runways
- Standardized first and second stage interfaces
- Standardized interfaces for strike, C2, ISR, mobility and spacelift payloads
- Operationally significant payload capacity to orbit (e.g., from DARPA XS-1 up to Evolved Expendable Launch Vehicle (EELV) class)

This full capability can potentially be demonstrated in as little as 5 years, and made operational in 7 years. Suborbital capabilities leveraging existing commercial platforms, and perhaps the DARPA XS-1, can be used to demonstrate future capabilities and gain important practical operational experience in the near-term.

FINDING [F.2]: Current Joint Requirements Oversight Council (JROC) requirements are not enabling to Fast Space: While technology exists to greatly enhance Joint Force freedom of action in the space domain and associated C2/ISR for terrestrial domains, our requirements have placed us in a strategic cul-de-sac. Despite over a decade for calls for re-usable launch and responsive launch on demand including by Congress and USSTRATCOM, a perception persists that the legacy Mission Needs Statement (MNS) remains valid. This problem—a lack of JROC-validated and documented requirements—was reported[5] by the DOD to the US Congress in June 2015 in response to the 2014 National Defense Authorization Act.[6]

This DOD study *"determined the DOD has no formal requirements for operationally responsive launch. Two Joint Requirements Oversight Council (JROC)-validated documents—the Space Support Mission Area (SSMA) Initial Capabilities Document (ICD) and the Rapidly Deployable Space (RDS) ICD—identify the warfighter's operational needs and capability gaps for spacelift or space launch, in addition to the original Evolved Expendable Launch Vehicle (EELV) Operations Requirements Document (ORD). These are the only validated combatant command requirements documents. Unfortunately, none define "responsive launch," and thus, the term can be*

---

[5] Operationally Responsive, Low-Cost Launch (ORLCL) Congressional Report, June 2015

[6] PL 113-66, § 915 (2013)

widely interpreted. Also, launch needs identified in the ICDs associated with responsiveness are limited to small launch capability. Finally, these documents suggest a distinction between two launch categories as "launch on schedule" or "launch-on-demand," but lack specificity. In other words, there are no clearly articulated and validated requirements for operationally responsive launch today."

The US Government Accountability Office reviewed[7] the DOD report and asked the DOD about the reason for the lack of requirements for responsive launch *"DOD officials told us that such requirements are premature without a validated need for responsive launch. Officials from the United States Strategic Command (USSTRATCOM) added that responsive launch needs cannot be well defined at this time due to uncertainties in the threat environment, and stated that DOD will validate future responsive launch requirements once it acquires new information from intelligence and defense studies presently underway. In lieu of a consolidated plan, the DOD report calls for reassessments of responsive launch needs and national security space program architectures, to help clarify requirements, and to take advantage of emerging responsive launch options."*

Meanwhile, the USAF FY2017 budget request for the Operationally Responsive Space Office states: *"United States Strategic Command has identified three needs as a result of dramatically increased demand and dependence on space capabilities as follows:*

 a. *To rapidly augment existing space capabilities when needed to expand operational capability.*

 b. *To rapidly reconstitute/replenish critical space capabilities to preserve 'continuity of operations' capability.*

 c. *To rapidly exploit and infuse space technological or operational innovations to increase US advantage.*

If the Joint Force wishes to leverage the emerging capabilities of launch on demand, development of large constellations of small, low-cost satellites, and rapid-reconstitution of the same, the USAF must provide leadership to help the Combatant Commands (COCOMs) articulate their emerging requirements, and provide stewardship and advocacy of these requirements through the JROC process.

Finding F.2 directly supports Recommendation R.2 located on page 34.

OBSERVATION [O.1]: The capabilities provided by RLVs do not neatly align in any one Core Function Lead Integrator (CFLI)

The capabilities provided by RLVs do not neatly align in any one Core Function Lead Integrator (CFLI). Private firms are developing game-changing technology that could alter the way in which the Air Force achieves global vigilance, power, and reach. Fast Space and ULCATS enables compelling military benefits across a spectrum that includes C2, ISR, CPGS, Ballistic Missile Defense, PNT augmentation, and SSA. However, no part of the USAF is responsible for all those capabilities. However, there is no single champion to bring these capabilities together in the strategy, planning, and programming

---

[7] GAO Assessment of DOD Responsive Launch Report, GAO-16-156R, October 29, 2015

process. Each Core Function Lead may pursue elements of this concept individually but there is no single organization responsible for pursuing the concept. C2 and Global Integrated ISR are the responsibilities of Air Combat Command (ACC), Conventional Prompt Global Strike (CPGS) is the responsibility of Global Strike Command, while PNT and SSA are the responsibility of Space Command. The fractionation of utility and benefits for Fast Space and ULCATS is a barrier to accelerating development of this strategic capability.

Industry Views

SUMMARY: US commercial industry generally agrees, but not without exception, that the path to ULCATS is development of two-stage-to-orbit (TSTO) RLV. Multiple credible US firms are planning and investing in the development of TSTO RLVs, using existing and near-term technologies. While there are at least a half-dozen serious US companies investing significant private risk capital in RLVs, the survey team did not identify any US companies investing significant private capital in an *expendable* launch capability that had the potential of achieving ULCATS.

FINDING [F.1.5]: Industry believes TSTO RLVs are technically achievable based on today's technology: Of those companies surveyed and interviewed, which included both traditional and new space firms, there was strong agreement that TSTO RLV demonstrations are technically achievable today.

FINDING [F.1.6]: Commercial methods are more economically affordable for developing RLVs: Industry generally reported, and our analysis supports, that the cost of developing and testing a TSTO midsized RLV ranges from ~$2 Billion with a clean sheet design to ~$500 Million for those companies who have already developed, or are developing, the required rocket engines and a reusable first stage.

FINDING [F.1.7]: The primary barrier to RLV development is the commercial business case: The primary barrier to 100% private development of an RLV is sufficient proven market-based demand to justify the large high-risk private investment.

FINDING [F.1.8]: Industry supports use of risk-sharing OTAs for commercial partnerships: There is broad agreement, interest, and support from US firms of all sizes in partnerships that use the DOD's Other Transaction Authority (OTA) to accelerate commercial development of RLVs. For example, NASA's use of OTAs to develop Commercial Orbital Transportation Services (COTS) has been particularly effective.

Technical Feasibility Assessment

SUMMARY: Multiple US companies, from traditional aerospace firms to billionaire-funded companies, are pursuing a broad range of technical strategies to achieve ULCATS. All systems to date are only partially reusable. However, several credible companies have plans to achieve full reusability. Our technical team has concluded their RLV development plans are technically feasible.

FINDING [F.1.9]: 1st Generation commercial fully reusable LVs are technically feasible: The first generation of fully reusable LV systems being planned by commercial industry are technically feasible. They have a solid foundation for meeting orbital performance requirements with advanced structures (aluminum lithium alloys and advanced composites) for low mass and rocket engines (e.g. Merlin 1D, BE-3 and BE-4, AR-1, etc.) having high performance, reliability, and low costs. Both SpaceX and Blue

Origin are actively developing orbital launch vehicles that have reusable first stages as the step prior to developing their fully reusable systems. In addition, DARPA's XS-1 Program has received proposals to fly a partially-reusable LV 10 times in 10 days that will lead to operational systems having significant cost reductions over present expendable systems.

FINDING [F.1.10]: 1st Generation full reusability is still a difficult technical challenge: While the benefits of full reusability are likely to be transformational and worth the risk, development of a fully reusable LV is not without significant technical challenges. Full reusability is significantly more challenging than the development of a 1st generation partially reusable LV. While the key technologies required are available today, the integrated capability of fully reusing a propulsive stage that goes all the way to orbit has not yet been demonstrated or attempted. In other words, the Technology Readiness Levels (TRL) are sufficient for 1st Generation fully-reusable LVs, but the Integration Readiness Level (IRL) is lower and where the true challenge lies. An extensive design, development, test and evaluation (DDT&E) effort will be required. Leadership needs to understand there will almost certainly be high-profile failures during the development process, just as there have been spectacular failures in developing reusable first stages. Future commercial development partnerships need to be designed with a robust mindset, culture, budget, and schedule that enable the companies to rapidly recover and learn from these likely failures on the path to success.

FINDING [F.1.11]: New technology will support the virtuous cycle to achieve improvements in aircraft-like operations: Consistent with a strategy for jump-starting a virtuous cycle (**Figure 2**) of continuous improvement, new advanced technologies will be required. New technologies beyond 1st generation fully-reusable LVs are required to achieve higher levels of reusability, reliability, operability and quicker turn-around in future generations of RLVs.

FINDING [F.1.12]: The systems integrators are in the best position to lead the technology prioritization process: There are numerous proposals for advanced RLV technologies that could be developed, but choosing the technologies with the highest value depends on the particular RLV system design they support. For this reason, the system integrators designing and developing the current commercial RLV systems will have the best insight on which technologies would have the most significant impact on improving next generation RLV systems to greater levels of reusability, reliability, operability and cost reduction.

Findings F.1.11 and F.1.12 directly support Recommendation R.5.1 and R.5.2 located on page 35.

Lowering Launch Costs

SUMMARY: Lowering launch costs by 10X to achieve ULCATS requires jump-starting a virtuous cycle (see **Figure 2**) of industry competition with commercial RLVs that open up new markets. Initially, costs can be reduced by 3X. As new markets and applications develop based on the availability of 3X lower launch costs, this will increase flight rates, and the availability and reliability of launch services. This will increase investor confidence, driving investments in the next generation of RLVs, with increased reliability, robustness, and operability, lowering costs and increasing flight rates even further. This cycle could achieve a 10X reduction in launch costs. See **Figure 1** in Executive Summary.

FINDING [F.1.13]: The development of multiple competing fully-reusable launch vehicles could lower prices by a factor of 3X in the near-term: Assuming competition among fully-reusable LVs and new market development, our analysis shows that launch prices could be reduced by a factor of 3X.

OBSERVATION [O.2]: An initiative to develop much lower cost launch vehicles may not be sufficient, on its own, to deliver a 10X reduction in launch costs: The development of fully-reusable LVs, on its own, will only achieve an estimated 3X reduction in launch costs. An effective ULCATS strategy will address cost drivers from all of the following:

- Barriers to driving up flight rate in new markets and in financing new systems
- Insuring sufficient levels of industry competition
- Policy, legal, and regulatory barriers
- Barriers to integrating current technology (partnership tools and methods), and developing future technology for the next generation of systems

OBSERVATION [O.3]: Key cost drivers are: (a) Flight Rate, (b) Reuse Rate, and (c) the Labor Intensity of Operations, and each of these factors are inter-related. Both high flight rate and high hardware reuse are needed to generate increases in reliability to achieve aviation-like labor efficiencies. High-flight rates are also necessary to increase the rate of lessons learned. Reusability tends to require larger mass margins that increase dry mass, which lowers the payload fraction. Therefore, without a decrease in the labor intensity of operations, the cost per pound for a launch will increase. However, our analysis shows the performance hit can be more than regained through hardware reuse and operational efficiencies enabled by increases in reliability, both leading to lower labor intensity of operations. Higher flight rates are required to regain the per flight profits lost from offering lower pricing, to attract the capital needed to fund RLV development, and future innovations.

OBSERVATION [O.4]: Labor Intensity of Operations is the most ignored key cost driver: Our analysis shows that labor intensity of operations is the most ignored and misunderstood cost driver as most expendable launch vehicle operators do not understand aircraft-like reusability operations. Developing a RLV without also redesigning how the RLV is operated yields much smaller cost-reduction benefits. Using this approach, our analysis shows that business cases with 3X price reductions could be supported at likely medium-term flight rate improvements assuming COTS-like partnerships with the USG. Analysis suggests even larger cost reductions are achievable if strategic investors or philanthrocapitalists choose to ignore amortization of nonrecurring cost (NRC) to gain market share and build longer-term value versus maximizing near-term return on investment. While not ULCATS, these prices and flight rates should jump-start a virtuous cycle of gaining operational cost efficiencies while moving toward aircraft-like operations. This will drive more demand and higher flight rates, leading to further price reductions and more demand, all factors driving towards ULCATS.

For example, holding Launch and Flight Readiness Reviews days before launch (a launch-by-launch certification process) is a visible sign of an outmoded set of methods that must be replaced. While needed for previous and existing systems that lack proof of dependable operations, the serial flight-by-flight process is incompatible with the system characteristics required of a responsive military capability and a thriving space economy. Achieving ULCATS will require replacing these methods with a one-time testing process to verify reliability prior to fielding an operational service. This type of approach will be far more compatible with the time-to-market and national security space needs for our nation's future.

FINDING [F.1.14]: Competition is required for large reductions in price in commercial markets: Reusability without competition will result in small, if any, price reductions to current customers of space launch. During the industry survey, multiple companies stated their plan was to capture the value of the cost reductions with higher profit margins. They indicated they would only pass on significant price reductions to their customers if forced to by competition.

FINDING [F.1.15]: Advances in reusability are likely to start a virtuous cycle: As illustrated in **Figure 2**, a critical element in achieving ULCATS is jump-starting a virtuous cycle of innovation, investment and market growth. When combined with competition, first generation RLVs designed for operability can lead to significant cost savings. Reusability could enable greater operational agility and responsiveness for more assured and frequent launch. Improved pricing and availability will allow more business plans to close, which will support more aggressive fleet sizes and deployment rates and create higher flight rates. This is particularly true for certain use-cases where demand is price-elastic—where reductions in price will drive larger increases in demand, such as space adventure travel. Lower prices and increased availability also encourage new spacecraft system designs and architectures that are not based on maximizing performance per pound. Instead, these designs take advantage of lower launch costs to build more massive spacecraft, optimizing features other than minimizing mass that result in even higher flight rates.

The benefits derived from reusability and aircraft-like operations will be passed on to end users as lower costs and/or higher quality services. Lower costs and higher quality services should drive demand for more space segment capacity and thus higher flight rates. These higher flight rates combined with reusability allow lessons learned to be engineered back into subsequent vehicle models. Better vehicle designs can generate additional operational cost savings through lower labor intensity for maintenance and refurbishments. The virtuous cycle starts all over again, as the process attracts increasing levels of private capital investment.

One example of a virtuous cycle is the early decades of aviation. A combination of American entrepreneurial innovation, intense industry competition to capture new commercial markets, and smart US Government investments as a partner in the 1920s and 1930s created a virtuous cycle that established America as the global leader in long-range aviation. In direct contrast to World War I, American leadership in aviation played a pivotal role in World War II. For those who care about national strategy, it is critical to understand why America recaptured leadership in aviation after losing it.

In the 1920s, American aviation was on a par with other leading countries in the world. During the '20s and '30s, the US Government actively stimulated commercial American aviation using a partnership-based strategy. The US Post Office purchased commercial airmail (Kelly Act of 1925), and the NACA assisted industry with technical challenges such as drag reduction, de-icing, and variable pitch propellers. Dozens of private US airlines were created in the 1920s, many of them aggressively targeting the passenger travel market. The existence of a long-range passenger market in a single country spanning a continent provided a critical market incentive to US airlines that drove commercial investment and innovation. In the early 1930s, US firms rapidly leapfrogged each other with introduction of new advanced planes like the Northrop Alpha (1931), the Boeing 247 (1933) and the Douglas DC-2 (1934). The London Morning Post reported *"America now has in hundreds, standard commercial aeroplanes with a higher top speed than the fastest aeroplane in regular service in the whole of the Royal Air Force"*.[8]

Then in 1936, Douglas introduced the DC-3, which had 50% more passenger seats than the DC-2, but cost only 10% more to operate. The DC-3 was a tipping point in the commercial aviation industry—airlines no longer needed airmail on a route to be profitable. They could make a profit solely on private passenger travel. It was so successful that four of every five airliners in the United States in 1941 was a DC-3[9], and over 10,000 DC-3s were built world-wide. However, it too was replaced in this virtuous cycle by the next generation of American aircraft, such as the Lockheed Constellation (1943) and the DC-6 (1946). As a result of this virtuous cycle that leveraged private investment and innovation, America had a dominant technological leadership position in large long-range aircraft in WWII.

FINDING [F.1.16]: In the Long-Term, a successful virtuous cycle can enable a launch price reduction of 10X: There is no fundamental physical or economic barrier to an order of magnitude reduction in launch costs—and even more if true aircraft-like operability is achieved. With a virtuous cycle of new markets, higher flight rates, more investment, spurring ever more advanced technologies, leading to reductions in the labor intensity of operations, a 10X reduction or more is reasonably achievable. In the long-term, the primary floor on launch costs could become the cost of propellant, or the non-propellant source of energy used for launch.

OBSERVATION [O.5]: Expendables are unlikely to achieve ULCATS: Some argue higher flight rates can also be served by expendables with volume production creating scale efficiencies leading to a virtuous cycle. The consensus of this team is that the advocates of expendables take this argument too far. Based on their logic, we would have expendable airplanes that you fly once and then throw away, because expendable airplanes would be so cheap to make. Launch vehicles are very complex, capital-intensive integrated systems of complex technologies, similar to airplanes, railroads, ships, trucks and automobiles. Launch vehicles are not mass-manufactured commodities like soda cans, matches, and bottles. There is no example in history where complex transportation systems became economical and affordable based on expendable systems. Reusability provides the only hope to achieve ULCATS.

ULCATS requires both hardware and operational cost savings. Operational efficiencies are hard to achieve without the safety, reliability, and lessons learned created by robust reusable systems. While the first flight of an airplane or RLV will receive intensive testing and verification, the tenth flight will have much less labor intensity, enabling rapid low-cost reflights. Further, there are real supply chain production cost limits and labor efficiency challenges required to support hundreds to thousands of annual expendable launches.

---

[8] Douglas J. Ingells, "The Plane That Changed the World: A Biography of the DC-3", p. 75

[9] David J. Cartwright, "Sky As Frontier", page 100

Market

FINDING [F.1.17]: Big LEO constellations are unlikely to close their business case in the near-term without RLVs: Many commercial companies are attempting to finance and develop large LEO constellations of small satellites to provide global ubiquitous broadband communications services. As a result of the revolution in low-cost small satellites and the potential to reduce costs even further using mass production techniques, the total cost of creating these constellations is now dominated by launch costs. Launch costs are approximately 1.5X to 3X the cost of satellite manufacturing for the space segment of these businesses. Traditional investors have not yet invested the capital needed by the large LEO constellations. Our business case analysis shows that the combination of low prices paid by global broadband customers, the lack of customers over 90% of the planet, and the continuing high-cost of launching satellites, makes it unlikely that traditional investors will invest the billions of dollars needed to finance these systems without significant launch cost reductions.

FINDING [F.1.18]: Current launch demand is not large enough to support a 10X reduction in launch costs: The global flight rate for medium to heavy launches is approximately 40 launches per year and many of these flights are not available due to subsidies and national security interests. These rates would not be sufficient to sustain multiple competing ULCATS systems, but there are promising new use cases that could provide significant increases in flight rate, especially as the cost of launch decreases.

FINDING [F.1.19]: New markets enabled by a 3X cost reduction can lead to ULCATS: We have identified three near-term and one long-term commercial use cases to drive flight rates to support 3X price reductions to achieve flight rates necessary for ULCATS. Our projection is that these markets could potentially add 100-200 flights per year. While 100 flights/year is sufficient to sustain 2 RLV firms, we believe these markets may not increase flight rates quickly enough on their own to attract sufficient levels of private capital to sustain at least 3 competitors. Early purchases of commercial services by one or more USG entities — perhaps from the LEO constellation service providers — may also be needed to provide critical evidence of sufficient market demand needed by some investors.

- <u>Big LEO Constellations</u>: Business plans exist for deploying 13,000 to 18,000 satellites in the 200- to 1,000-pound class (e.g. SpaceX, Boeing, OneWeb, etc.). A baseline of 9,000 satellites (50% of the maximum proposed) deployed over a six-year period results in roughly 1,500 satellites per year. These deployments are believed to use mostly medium to heavy launchers. Assuming an average of 30 satellites per launch yields roughly 50 flights per year.

    There is a synergistic relationship between ULCATS and big LEO constellations. While these constellations help close the business case for investments in ULCATS systems, ULCATS significantly improves the business case for large constellations. Launch cost is currently the single dominant cost of developing large constellations. Our economic analysis shows these systems are unlikely to close their business cases and raise the large amounts of capital required without much lower cost launch, and/or advanced purchase commitments by the USG for their services.

- Propellant for In-space Transportation: There is growing interest in delivering propellant to support civil space exploration and commercial satellite in-space fueling strategies. Companies are proposing many solutions, including launch from Earth and mining propellant from the Moon and asteroids. The amount of propellant in LEO required to support a human mission to Mars, based on NASA's Design Reference Architecture 5.0, is estimated to exceed 1.4 million pounds. Assuming launches from Earth to support one human mission to Mars every 26 months, this would require roughly 30 medium-lift flights to LEO from Earth every year. When including LEO propellant deliveries to transport commercial satellites travelling to Geosynchronous Earth Orbit (GEO), the flight rate for just propellant delivery could rise to 35+ per year.

- Space Adventure Travel: A few companies we interviewed are focused on this market and have invested considerable funds. They expressed the intent to invest significantly more capital in the near term. Initial systems will provide sub-orbital flights or carry small numbers of passengers to LEO, but long-term plans are more aggressively focused on very large numbers of passengers, high flight rates, and destinations beyond LEO.

  Of the commercial use cases we evaluated, this one is the furthest along in attracting private capital and developing RLV systems. Of the near-term markets, this has the greatest potential for driving higher flight rates and transforming how we think about and use space.

  There is significant uncertainty on how rapidly this market will develop. Two surveys were conducted in the early-to-mid 1990's to assess the likelihood of people buying rides to orbit at varying price points. Those surveys indicated a very high level of price elasticity for adventure space travel. Updating those results for 2016 dollars and adjusting for the higher number of wealthy individuals today yields global annual passenger rates of approximately 500 at $1 million per ride, 10,000 at $500,000, 200,000 at $100,000, and 500,000 at prices under $50,000. Assuming development of a large RLV, which transports 100 passengers per flight to LEO, this yields 5 to 5,000 flights per year, or roughly 100 flights per week at the lower price per seat range.

  If a large 100-seat RLV charged $50 Million per flight, this would equate to $500k per seat. The market survey suggests this kind of RLV could achieve a flight rate of 100 flights per year if it could keep up with demand. The projected revenue potential of such an RLV would be $5 Billion per year. While this kind of system would almost certainly need a destination to drive this level of demand, some of the systems being discussed by existing commercial firms have this potential.

- Space Solar Power (SSP): We believe there is potential for SSP to acquire sufficient financial and government support to fund an SSP technology demonstration project, in part to develop the national capability to assemble larger structures in space. Sophisticated industry participants like Northrop Grumman and CalTech[10] are devoting material capital and human resources to SSP technology development, as are countries like China[11] and Japan[12]. A successful demonstration could lead to pilot systems of 10-150 Megawatts that might provide services for in-space power, price-insensitive military needs, remote communities, and emergency disaster response. A technology demonstration or pilot system could require 10 – 100 flights over a 1-2 year period depending on its size and mass.

Much larger markets for clean electricity exist for baseload power, which would justify development of larger SSP systems in the Gigawatt class. However, those systems are unlikely to be enabled by an initial 3X reduction in launch costs from first generation RLVs, so it was beyond the scope of our study to assess them. In addition to first generation RLVs, these larger systems are likely to require in-situ resource development from the Moon and asteroids, advances in automated manufacturing, on-orbit construction, and satellite servicing. The required capabilities are likely to be within reach if smaller SSP prototypes are developed and a virtuous economic cycle in space transportation is jump-started. After US industry develops and demonstrates the first generation of fully-reusable LVs, the USG would be reasonably well-positioned to consider development of an SSP pilot demonstration system.

Financial

SUMMARY: Philanthrocapitalists and strategic investors are the most likely partners to co-invest with the USG in ULCATS systems. Philanthrocapitalists are significantly motivated by trying to make a difference for humanity and are willing to accept a higher degree of risk than traditional investors. Strategic investors, primarily major aerospace companies, view launch as strategically critical to future business and are also willing to accept a higher degree of risk.

OBSERVATION [O.6]: Traditional investors are not likely sources of capital for first generation RLVs: The vast majority of current launch investments have been made by governments and the aerospace industry. Financial investors like venture capitalists, private equity firms, and public capital sources have largely ignored this sector. This industry is capital-intense, has long up-front periods of negative cash flow, is regulated, has highly-competitive and subsidized markets, traditionally low profit margins, lumpy contract orders, and general lack of growth in annual launch rate.

Recently, there have been reports of over $2 billion of venture money flowing into the "NewSpace" sector. This statistic, however, can be misleading. One billion of that

---

[10] "Space-Based Solar Power Project Funded", https://www.caltech.edu/news/space-based-solar-power-project-funded-46644, 2015-04-28

[11] "China sets up laboratory to research building solar power station in space", http://www.scmp.com/news/china/policies-politics/article/1922390/china-sets-laboratory-research-building-solar-power, 2016-03-08

[12] "Japan Demoes Wireless Power Transmission for Space-Based Solar Farms", http://spectrum.ieee.org/energywise/green-tech/solar/japan-demoes-wireless-power-transmission-for-spacebased-solar-farms, 2015-03-16

investment came from the Google founders and Fidelity investing in SpaceX, and $500 million was from strategic investors in OneWeb. Of the venture capital funding, roughly $360 million went into three microsat remote sensing and weather businesses: SkyBox Imaging (now Terra Bella), Planet Labs (now Planet) and Spire. Tens of millions of dollars have gone to support a few new small launch ventures like Rocket Labs and Firefly. The study team believes that these small launchers will primarily serve the low revenue niche cubesat markets and the small volume replenishment market for LEO constellations.

OBSERVATION [O.7]: Philanthrocapitalists and strategic investors will lead: The primary candidates to serve as partners with the USG are the major industry participants, including aerospace primes and several newer companies founded and supported by space enthusiast billionaires. These industry players have the deep pockets, longer investment horizons, technical competence, human resources, and strategic visions to act as effective lead partners. Greenfield start-ups would face a large competitive disadvantage. However, in later years, once predictable revenues and cash flows are evident, further growth could be funded by private equity firms and public capital markets. While these later phases of a high-growth virtuous cycle may also require government support, it is too early to conclude what that support might entail. It is also quite possible in later years there will be a bifurcation of launch vehicle manufacturing and launch service operators just as occurred in the aviation industry.

Partnership Strategy

> **SUMMARY:** The USG's traditional acquisition methods are unlikely to achieve ULCATS. Non-traditional partnerships using OTAs have a much higher chance of success. The USAF has the existing authorities it needs for non-traditional partnerships to jump start the virtuous cycle with commercial firms.

FINDING [F.1.20]: OTA-based commercial risk-sharing partnerships are proving successful: The USG has repeatedly used OTA agreements to create unique commercially-led public-private partnerships (PPPs) to spur the development of new space capabilities such as DOD's EELV program and NASA's Commercial Orbital Transportation Services (COTS) program. In fact, OTA-based partnerships are the only acquisition methods that have successfully developed a new US operational launch vehicle since the Space Shuttle program nearly 40 years ago.

Beyond the development of launch vehicles, the DOD has used OTAs successfully to rapidly develop initial prototypes at low-cost; examples include the Arsenal Ship, Advanced Short Take-off Vertical Landing, Joint Unmanned Combat Air Systems, and Global Hawk. While some of these systems — such as EELVs and Global Hawk — became significantly more expensive over time, their cost growth was driven by traditional FAR-based approaches, imposition of additional requirements, and a lack of commercial markets to share costs and to create competitive pricing pressure.

FINDING [F.1.21]: Commercial OTA partnerships are much lower cost than traditional FAR-based methods: The NASA COTS program demonstrated a factor of 8x reduction in the development cost of the Falcon 9 launch vehicle between the actual costs and what had been estimated with NASA Air Force Cost Model (NAFCOM).[13] An order of magnitude cost reduction — from the $40 Billion previously estimated[14] by NASA and the USAF — is achievable for development of ULCATS systems.

FINDING [F.1.22]: The DOD has the OTA Authority (10 USC 2371b) it needs for ULCATS partnerships: The DOD has a long history of using its existing OTA authority well beyond the development of the Atlas V and Delta IV in the EELV program. The DOD's prototype OT authority is particularly well suited to accelerate the development and demonstration of commercial space systems. This authority has special provisions that allow follow-on production contracts or transactions, a key element in an effective partnership strategy.

FINDING [F.1.23]: A portfolio approach to commercial partnerships, on both demand and supply sides, will lower overall strategic program risk and has the highest likelihood of success: The results of a detailed analysis in this study indicate that a commercial development strategy — modeled after the lessons learned from NASA's COTS program, and the DOD's commercial OTAs to develop the EELVs — is well positioned as the catalyst needed to provide transformational breakthroughs in significantly lowering the cost of access to space. The cost reductions available from commercial partnership methods make it possible to afford a portfolio of partnerships. Further, industry firms clearly communicated that early purchases of services by the USG would significantly reduce industry's perceived investment risk and would increase the likelihood and size of investment. For these reasons, a portfolio of multiple commercial partnerships with commercial-style action on both demand and supply sides has the highest likelihood of success.

National Security Strategy & Policy

SUMMARY: Current international space law and existing space treaties are flexible enough to allow military usage of Outer Space. Further, it is in the long-term national security interests of the United States to lead in space development. This would allow the United States to establish key precedents in international common law, such as Western principles of free trade and commerce on the space frontier.

FINDING [F.3.1]: International space law, including the Outer Space Treaty of 1967, is flexible enough to allow for military usage of outer space short of deploying nuclear weapons and weapons of mass destruction or actively interfering with the activities of others: The conduct of all nations in outer space, as well as their citizens, is governed by international space law, which is itself a mix of customary international law (practices of countries), treaties and conventions, and bilateral or multilateral agreements between countries. The Outer Space Treaty of 1967 (OST) is the foundational treaty of outer space law, and contains only a few explicit prohibitions that might impede the Air Force's access to and use of outer space. Article IV prohibits the deployment "in orbit around the Earth [of] any objects carrying nuclear weapons or any other kinds of weapons of mass destruction, install[ing] such weapons on celestial bodies, or station[ing] such weapons in outer space in any other manner." The treaty is silent as to

---

[13] https://www.nasa.gov/pdf/586023main_8-3-11_NAFCOM.pdf

[14] NASA-USAF 120-day study, 2002

what constitutes a "weapon of mass destruction," but the United States position[15] is that weapons of mass destruction must be of "the same type" as to result in a "catastrophe that a nuclear weapon would lead to." Article IV also prohibits the establishment of a military base on the Moon or other celestial body, but does not prohibit the use of military personnel in peaceful exploration of the Moon or other celestial bodies.

Article IX contains an explicit prohibition against any nation engaging in activities that "would cause potentially harmful interference with the activities" of others. The United States has interpreted the term "interference" broadly, including the jamming of communications satellites, and would certainly include any physical harm to any US asset in orbit.

FINDING [F.3.2]: Because much of international space law is based on customary international law, whoever leads in outer space development is likely to establish operational legal principles, values, and practices on the space frontier that will last far into the future: Notwithstanding the reference to the OST above, many of the fundamental principles of outer space law were created through the unilateral and bilateral actions of individual countries, which once unopposed by the international community, have become customary international law. The concept of "free overflight" was established by the flight of Sputnik I in 1957. All countries are free to place vehicles in Earth orbits that fly over the territories of other nations, because no country objected to overflight by the satellite. The world came to accept as customary international law the concept of free overflight.

Similarly, the fact that no country formally objected to the United States or Soviet claims of a property right in returned Moon samples represents strong international customary law that pieces of non-man-made space objects returned to Earth become property. Finally, an Apollo 11 astronaut exercised a fundamental personal right of freedom of religion by serving himself communion on the lunar surface shortly after landing.

These examples demonstrate that the "first movers" in space development will be in a position to impact the development of international space law through their activities. The United States will be in a position to extend its notion of property rights, individual liberty and freedom, and other core concepts of Western civilization and culture, only if the United States takes the lead in developing outer space.

FINDING [F.3.3]: It is in the national security interests of the United States to establish the Western principles of free trade and commerce, including free enterprise development and use of space resources, in international common law: If the United States abdicates its leadership role and does not move out beyond the surface of the Earth, Western principles might be displaced in space by the doctrines of our adversaries. A country could deposit a piece of an asteroid with the United Nations and declare that the only way to comply with the notion of "the benefit of all mankind" in OST Article I would be to treat all extracted resources from celestial objects as being "the common heritage of mankind." This would subject space resources to confiscation and reallocation

---

[15] Hearings Before the Committee on Foreign Relations, United States Senate, 90th CONG., 1ST SESS., at 23 (Mar. 7, 13, and Apr. 12, 1967). 71, Colloquy between Ambassador Goldberg and Sen. Gore, "electrical jamming" would be included in the type of "interference" prohibited under Article IX and subject to "diplomatic intercourse"); *Id.* at 75, Colloquy between Ambassador Goldberg and Sen. Morse, Article IX prohibits "any other type of interference" including physical damage.

to all nations of the Earth. From there, it would be an easy argument that a SSP satellite could be required to share the power downlinked to all countries, regardless of their ability to pay. The United States has fought such notions in rejecting the 1979 Moon Treaty. Without continued leadership and the ability to create customary international law through its first mover actions, the national security interests of the United States could be compromised.

Findings F.3.1 through F.3.3 directly support Recommendation R.3 located on page 34, and Recommendations 6 and 7 as detailed in Appendix C on page C-2.

National Leadership, Economic, and Soft Power Benefits

> **SUMMARY:** The benefits of ULCATS go well beyond utility to the warfighter, and achieving ULCATS will require leadership and action at the national level. ULCATS cuts across all space agendas, all space agencies, and all space programs.

"In peace, [strategy]...may gain its most decisive victories by occupying...excellent positions which would perhaps hardly be got by war." --Alfred Thayer Mahan

"[Our] responsibility is to seek it under the most advantageous circumstance in order to produce the most profitable result. Hence his true aim is not so much to seek battle as to seek a strategic situation so advantageous that if it does not of itself produce the decision, its continuation by a battle is sure to achieve this." –B.H. Liddell Hart

What role should the military play in deliberately strengthening the nation's industrial base? Some argue that the military should be an enthusiastic partner in the development of new markets, as economic might and wealth contribute to security and military power. Another view argues that it is not the DOD's role to develop the economy, that the military should confine itself to its assigned job of warfighting and leave economic policy to other departments. There is an intermediate position—one usually taken by the Navy in defending their ports and ship-building infrastructure. A military can ensure its ability to win and prevent wars by ensuring its freedom of action and superiority of position as a direct consequence of a nation's industrial base. In a very real sense, our ability to have a competitive advantage in hypersonic strike and space access directly depends on the capability of our space industrial base.

In the 19th century, Admiral Alfred T. Mahan articulated the interactions between naval power and maritime commerce, and sea power's special significance during times of peace. The strategic linkages between space commerce and space power are similar. It was British commercial maritime leadership and innovation that enabled Britain to build the most powerful naval fleet in the world. In the 21st Century, space economic power will extend America's ability to project power during times of peace. ULCATS will enable America to deploy and project power anywhere, at any time, with significant effects that may only be perceptible only over time.

It matters greatly whether or not we have an industrial base capable of developing airplane-like access to space. The economic growth of the nation of capturing new markets is a happy consequence. Because military freedom of action is tied to cost, anything that reduces the long-term direct costs to the military increases military freedom of action. New markets create more opportunities for America to access and innovate in space, increasing the speed of our innovation cycle. New markets allow the militarily relevant industrial base to become self-sustaining, reducing the burden on the DOD budget. In particular, if we capture new markets that create scaling effects from high flight rates and a virtuous cycle, it lowers the cost of products and services in the enterprise allowing the DOD to purchase more capability for the same cost. If another nation captures those same markets, the same benefits accrue to their military, instead of ours. To the strategic military actor seeking to set up a competitive strategic position, the conscious nurturing of an industrial base is a legitimate and important function.

The USAF should view an investment to accelerate development of commercial RLVs leading to ULCATS, and large commercial LEO constellations that leverage an ULCATS strategy, as an industrial base strategy to protect its freedom of action. A Mahanian-based theory of space power, as applied to ULCATS, would lead us to the following conclusions:

- America's entrepreneurial talent for creating wealth is now creating and enabling a transformation in American space power.
- Space power is a protector of democratic freedom.
- Space power is needed to protect the wealth created by space commerce, and the combination of both can impose a political order friendly to western values.
- By leveraging the American approach to developing commerce into the space environment, we will create a strategic situation in which the United States is likely to gain and hold the upper hand.
- A side benefit of a successful commercial competition in space is an industrial base that provides America with the most advanced space military equipment.
- Space power is a national power multiplier, not through military competition, but through the soft political leverage attained through commercial development.

Our current approach to power projection is no longer economically practicable. Thus, more than any other military branch, the USAF can serve as a *preventative* force during times of peace, to reassure friends of our support, help us gain new friends, and dissuade states from attempting to contest our leadership in space.

FINDING [F.3.4]: Achieving ULCATS will require national-level leadership beyond DOD: As specified in **Finding F.5,** on page C-2, the flight rates and market demand required to achieve a 3X price reduction, and the ability to jump-start a virtuous cycle to continue growth towards 10X, are not possible without a significant level of change in organizational mission, strategy, structure, and funding in other federal agencies (e.g., FAA, FCC, NOAA). The breadth and magnitude of institutional reform can only be achieved with national-level leadership, because it extends far beyond the mission and authorities of the DOD and its interdepartmental relationships. Fortunately, such leadership is justified by the broad public benefits *beyond national security* of ULCATS as detailed in **Finding F.3.5**.

***FINDING [F.3.5]:* ULCATS will create economic and diplomatic soft power benefits, which also contribute to US national security.** Economic power and other forms of "soft power" are critical long-run components of national security. A country's economic wealth establishes a country's ability to invest in national security capabilities. The United States needs to leverage "soft power" means to influence world events to our advantage. Former SECDEF Robert Gates spoke[16] to the value of soft power for national security:

> *"We can expect that asymmetric warfare will be the mainstay of the contemporary battlefield for some time...But these new threats also require our government to operate as a whole differently -- to act with unity, agility, and creativity. And they will require considerably more resources devoted to America's non-military instruments of power."*

ULCATS will deliver the following economic and diplomatic soft power benefits, which also contribute to US national security:

- Accelerated development of new markets and industries, and the creation of many tens-of-thousands of American jobs, as America uses our first-mover advantage to capture global leadership in many new 21st century industries. The current $300 billion per year space marketplace could grow to a trillion-dollar per year industry.

- Significant reductions in the cost of Internet access. US leadership will be maintained in this important industry by making access more affordable and available for billions around the world who are now offline. While several US companies plan to develop broadband satellite constellations, our economic and financial analysis demonstrates their business cases are unlikely to close without a 2X to 3X reduction in launch costs.

- Persistent and more accurate monitoring of Earth's entire environment at the local level on a 24-7-365 basis. This will create more accurate weather predictions and much better storm warnings, providing operational advantages for US forces, and at the same time benefiting the lives of billions.

- Weekly then daily launches to space, allowing thousands of private citizens to travel to orbit every year, which will advance US leadership in this important arena of international competition.

- The combination of frequent human trips to space into low Earth orbit, and more affordable easier trips by humans into deep space will deliver significant soft power benefits as the entire world is awed by American leadership, ingenuity, and entrepreneurship.

Findings F.3.4 and F.3.5 directly support Recommendation R.3 located on page 34 and detailed in Appendix C.

Purpose-Built Development Organization

FINDING [F.4] Traditional USG acquisition methods, or a traditional operationally-focused acquisition office, are likely to fail at effectively partnering with commercial space industry.

The majority of leading US commercial firms have made it very clear they are not interested in traditional USG FAR-contract-based approaches to accelerate private development of ULCATS systems. Traditional USG methods of buying launch services

---

[16] Robert M. Gates, Secretary of Defense, Landon Lecture Remarks 26 November 2007, https://www.k-state.edu/media/newsreleases/landonlect/gatestext1107.html, September 15, 2016.

have been optimized for removing residual levels of risk, not for lowering costs. The current USG methods of mission assurance are completely rational in an industry where the cost of the spacecraft is several times the cost of the launch and the consequences of a failed launch can be catastrophic to national security. USG agencies that implement these important methods of mission assurance have cultures, processes, and values that are in complete alignment with this philosophy of space launch.

These same processes, cultures and values — which are critical to these agencies' ability to eliminate the residual risk from expendable launch vehicles — are showstopping-barriers to the commercial innovation process. **Any ULCATS initiative that proposes to leverage commercial innovation will fail if it is given to a USG agency that develops or acquires systems using traditional governmental methods, or processes, or has a traditional USG development or acquisition culture**.

Finding F.4 directly supports Recommendation R.4 located on page 34.

Overall Recommendations

RECOMMENDATION [R.1]: Partner with US commercial firms pursuing ultra low cost access to space (ULCATS) using the DOD's Other Transaction Authority.

The USAF should assemble a team to pursue the *authority to proceed* with a competition for jointly-funded (cost-shared) prototype OTAs. The full and open competition will seek multiple US commercial partners to develop and demonstrate their proposed space systems in collaboration with USAF financial assistance and broader USG technical resources. As enabled by DOD's OT authority, follow-on contracts or transactions should legally allow for follow-on offers to the selected partners contingent upon the successful completion of the prototype demonstrations.

RECOMMENDATION [R.2]: Integrate consideration of Fast Space and RLVs into the Joint requirements and acquisition process: The current process of requirements and acquisition does not incentivize building ground-breaking capabilities. We recommend that relevant DOD organizations create initial capability documents (ICDs) that capture the full set of opportunities provided by highly-reliable RLVs that enable rapid-turn around and surge-launch on demand, and associated on-orbit capabilities, and champion these to the Joint Requirements Oversight Council (JROC).

RECOMMENDATION [R.3]: Shape the interagency environment. To maximize the operational and strategic utility of ULCATS, the SECAF should adopt a proactive approach to help shape national policy: As the Principal DOD Space Advisor (PDSA), the Secretary of the US Air Force (SECAF) has a broad view of how the alignment of civil, commercial, and national security can benefit comprehensive national power. We recommend the SECAF as PDSA take an active stance in maturing the policy and regulatory environment outside the DOD that can maximize the benefit of high launch rate commercial RLVs and associated on-orbit capabilities.

Such advocacy is likely necessary to achieve the strategic economic benefits discussed in this paper. A full list of recommended policy positions the SECAF may wish to advocate are contained in **Appendix C**. These include: proposing that the White House actively manage ULCATS-related reforms across the whole of the U.S. Government, advocating for restructure of commercial launch and spacecraft licensing, and proactive shaping of international norms.

RECOMMENDATION [R.4]: Create a purpose-built organization to manage commercial ULCATS efforts.

To succeed, the USAF needs to create a purpose-built organization to use innovative acquisition processes and methods, notionally called the "NewSpace Development Office" (NSDO).

This organization requires a "Fail-Fast, Fail-Forward" culture as opposed to the traditional operationally-focused risk-averse culture where "failure is not an option." Silicon Valley and NewSpace have proven the powerful advantages of faster innovation cultures that expect and encourage incremental tactical failures as a key part of their strategy for developing new systems and technologies. In order for the USAF to leverage the power of the NewSpace innovation culture, this USAF organization needs to understand, appreciate, and support a culture that is foreign to traditional USAF acquisition values and practices. Dr. Clayton Christiansen describes[17] the challenge as follows:

"A surprising number of innovations fail not because of some fatal technological flaw or because the market is not ready. They fail because responsibility to build these businesses is given to managers or organizations whose capabilities aren't up to the task. Corporate executives make this mistake because most often the very skills that propel an organization to succeed in sustaining circumstances systematically bungle the best ideas for disruptive growth. An organization's capabilities become its disabilities when disruption is afoot."

The organization must be lean, consisting of approximately a dozen people with a clear mandate. Because the function of the organization is to negotiate milestones and maintain oversight of achievement of milestones—as opposed to attempting to manage or control industry-internal processes—only a small organization is required. To benefit from the speed industry can provide, the organization must take a posture of cultivating trust based on aligned objectives, seeking insight rather than oversight of day-to-day industry-internal processes. The organization must be located in the communities where the commercial innovators reside.

The organization must be able to rapidly acquire and maintain both military and civilian expertise and therefore requires specific personnel authorities (detailed in **Appendix B**), including: Coding military billets as a Joint Duty Assignments, authority to extend military tours up to 5 years in order to establish continuity and trust, and civilian hiring authorities including HQE, IPA, PMF, and Section 1101.

The organization requires a sufficient budget to fulfill its mandate (including rapid prototyping, engineering demonstrations, fly-offs), and direct control of their own budget. The organization requires specific legal and procurement authorities to enable rapid RDT&E often referred to as "non-traditional or innovative acquisition." These include authorities to: employ Other Transaction Authority (OTA) for basic, applied and advanced research; prototype project authority; purchase for experimental purposes; and the authority to award incentive prizes.

The organization requires a strong leader who embraces these cultural elements, is committed to the purpose, and has a proven track record delivering results. DOD

---

[17] Dr. Clayton Christiansen, "The Innovator's Solution", Chapter 7, page 177.

precedents for similar purpose-built organizations include the Naval Reactors program led by Admiral Hyman Rickover and the ICBM program led by Gen Bernard Schriever. A key ingredient for success in such organizations has been direct access to senior leadership with few additional oversight mechanisms.

The recommended plan to set up the organization is provided in **Appendix B — Purpose-Built Organization Implementation Plan.**

RECOMMENDATION [R.5.1]: In partnership with commercial industry, develop a prioritized investment list of next generation RLV technologies: The NSDO should periodically survey private industry to identify and prioritize the most important RLV technologies that the USG should develop for the next generation of RLVs. This industry-led process must be open and transparent; it could be managed by the "commercial advisory council" of the NSDO.

RECOMMENDATION [R.5.2]: Invest in research and development at NASA and the Air Force Research Laboratory (AFRL) according to commercial industry's prioritized list of RLV technologies: The list of next generation RLV technologies, based on the prioritized inputs from commercial industry, should be delivered by the responsible organization to NASA and the AFRL, which are the nation's leading space technology advanced research and development organizations.

Conclusion

The USAF stands at a moment of opportunity. Due to the innovations of our private sector, we are briefly ahead of our strategic competition. As Airmen who seek advantage in the space domain, we must champion this opportunity to build an enduring architecture of advantage. A commanding lead in RLVs or any approach to Ultra Low-Cost Access to Space (ULCATS) constitutes an important counter-move to the A2/AD strategies of our adversaries.

The USAF has agency to make a bold move. The door is open with industry to begin OTA-based public-private partnerships that could deliver a menu of capabilities for Air Force and Joint functions within five years. Once that decision is made, three lines of effort must be pursued. The first is to establish a purpose-built organization designed to partner with industry and accelerate the development of these capabilities. The second is to take an active role shaping the interagency and commercial policy environment. The third is to take an active approach to writing and championing requirements documents for RLV systems with rapid turn-around, and surge launch-on-demand capabilities that are economically affordable to develop and sustain.

If the Air Force takes these steps, it increases its own freedom of action with a new vector for global vigilance, reach, and power projection. But it does even more for the nation—it puts the United States on an industrial learning curve that opens new economic vistas for long-term economic power. Once ULCATS exists, we will have opened up a virtuous cycle where higher launch rates lead to ever-lower marginal launch costs, and lower launch costs lead to yet higher launch rates.

Our strong recommendation to Air Force leaders is not to let this moment in time pass. Delayed action runs the risk of losing America's lead to a fast follower. The industry consensus is that ULCATS creates the opportunity for humanity to access what may become multi-trillion-dollar industries in the space domain. A failure to see this as a strategic industry—like ocean-going ships or passenger aircraft—risks losing significant national security benefits outside the USAF; the loss of significant market share in strategically critical space industries; the loss of the carrying trade to a 3rd dimension; and the loss of US-flagged carriers for global broadband and remote sensing. A failure to partner early with industry slows the USAF OODA loop to apply these technologies for C2, ISR, mobility, and power projection even while our adversary is taking these steps. On the other hand, the dream of Airmen since the early 1960s of fully reusable, sortie-access to space has never been as available in time or cost. We need only walk through the door that is open before us.

Appendix A — Frequently Asked Questions

ULCATS needs the Big LEO constellations, but why do the Big LEO constellations need ULCATS? Why do we need ULCATS to get Big LEO C2/ISR solutions?

Providing low-cost affordable Internet access to "the other 3 billion" is a compelling idea that motivates many space social entrepreneurs. This was the vision of Craig McCaw and Bill Gates when they started Teledesic in 1994.

Unfortunately, the business case for the Big LEO constellations still does not close. Sophisticated investors remember the promise and the bankruptcies of Iridium, Globalstar and ORBCOMM, and the collapse of the Teledesic, Skybridge and ICO ventures. Sophisticated investors know the real reason 3 billion people still don't have Internet access is because these people can't afford to pay enough to justify a commercial investment to serve their needs. This is the reason that no large investor, including Google and Facebook, has financed a LEO constellation broadband system.

However, if the cost of providing Internet access can be radically reduced, the business case could be closed. The only way to radically reduce the costs of developing large constellations in LEO is by achieving ULCATS.

OneWeb is an illustrative example of the problem. They plan to mass-manufacture 840 spacecraft on an automated production line for $500,000 each. OneWeb plans to launch 32-36 spacecraft at a time using the Soyuz launch vehicle. Assuming ~$50 million per launch, that is $1.4-1.5 million per launch for each spacecraft. Therefore, OneWeb's ratio of spacecraft launch cost to production cost is about 3 to 1. Approximately 75% of the total cost of building out the OneWeb space segment is in the launch costs.

To achieve the vision of global ubiquitous affordable Internet for everybody, and to implement the vision of global ubiquitous dynamic C2/ISR, we need ULCATS.

We have tried this before, and all previous attempts failed. How is this different?

Many things are new since the previous failed attempts to achieve ULCATS.

- Emerging threats, both in orbit and A2/AD, make it critical that we take a fresh look to find solutions to our most important national security challenges.

- In previous decades, immature technology was a primary inhibitor. Today, technology is sufficiently advanced to pursue demonstrations of fully-reusable LVs.

- Today's efforts are led by billionaire industrialists who are investing their own money. The fact that industry only financially gains if it "succeeds", and not just for "effort", is a significant difference.

- Previous attempts were government-led and -dominated activities that did not leverage the innovation of private sector entrepreneurs. The recent very public successes by billionaire industrialists in reusable launch demonstrate that commercial industry is on a path to success.

- Previous attempts have been focused on overcoming the technology barrier.

These advancements have largely avoided transforming the operational army that drives up operations costs between each flight. By validating the reliability of ULCATS systems for many flights during initial testing, and by investing in process innovations leading to rapid turn-around by smaller teams, the labor intensity of operations will be transformed.

- We now have convincing proof OTA agreements — creating true partnerships with commercial firms — work at much lower cost than traditional approaches.

Didn't the Space Shuttle prove that RLVs don't work?

The Space Shuttle never achieved the status of being a "reusable" launch vehicle—it was a partially-reusable, partially-refurbishable, partially-expendable launch vehicle. There are two primary lessons learned from the Space Shuttle experience that are of strategic relevance to an ULCATS strategy:

- The technology was not ready in the 1970s to support successful development of fully reusable LVs.

- A government agency, even a well-run agency, does not have the correct economic incentives to lower costs. NASA was not incentivized to eliminate the operational and labor costs that were part of the Shuttle system. Since NASA depends on political support, which is driven by the number of jobs in congressional districts, the opposite is true.

America has proven over its history that private industry, properly incentivized in a pro-competition environment, is much more successful at lowering costs. As a result it is private industry that designs our automobiles, airplanes, trains and general transportation systems, while the USG serves more effectively as a customer, regulator, and investor in next generation technologies.

Didn't the X-33/VentureStar prove RLVs don't work?

There are many lessons learned of strategic relevance from the X-33/VentureStar:

- Lockheed had no significant economic incentive to make the X-33/VentureStar work. Lockheed made money on both Titan IVs and the Atlas V, and so it was in part competing with its own economic interests.

- While Lockheed put some government-funded IR&D into X-33, they did not put any real significant investment of corporate capital into the venture. In other words, Lockheed did not have any real "skin" in the game.

- When NASA decided to pick only one winner this eliminated the potential incentive for competitive pressure to motivate Lockheed to make the program work. By winning the bid phase, Lockheed had already achieved a partial victory by eliminating the competition to Titan IV and Atlas V.

- There was a disconnect from the beginning in how the X-33 program was structured. For some it was an X-vehicle with the purpose of flight demonstrating advanced technology — as illustrated by the program's name. For others, it was a Y-prototype vehicle that served as the critical next step to Shuttle replacement. If it was the former, an X-vehicle, it should have been limited to the purpose of demonstrating technology, and asking industry for risk capital was not appropriate. If it was the latter, a Y-prototype vehicle leading directly to an operational profit-making business, a large industry co-investment of real private capital was needed to insure an appropriate alignment of incentives.

- NASA chose the highest technical risk solution from among the three major X-33 bids, as they gave higher points to the bidders with the "most new technology". This drove the technology risk of the X-33 program up higher than necessary if it was supposed to be a prototype leading to an operational vehicle. This is the exact opposite of how almost any commercial firm would have evaluated the process, as "technical risk" is something to be eliminated and mitigated before taking a product or service to market.

The National Aerospace Initiative (NAI) was estimated to cost $40 Billion in 2003. How are you so much cheaper?

There are a couple reasons the NAI was so expensive as compared to the cost of commercial approaches to developing RLVs.

- Commercial development processes are inherently lower cost than traditional methods — and can be as much as 8-10 cheaper in some cases — which was proven by the NASA COTS program.
- The NAI had to make everybody happy, so it included a significant piece of work for all parties. The "political process" required to achieve buy-in and consensus significantly increased the estimated price.

The traditional business case for an RLV does not close. So, why should we do this?

We agree that the pure commercial business case — based on the traditional risk-adjusted return on investment (ROI) — does not close. The early days of aviation (air mail) and railroads (transcontinental railroad) had the same problem. In all these cases, the benefits to national security, and to the public and society were so large, that it justified government action to accelerate private development.

Further, if the traditional business case closed for an RLV, the USG would not need to act. Private investors would fund the development of RLVs. We have a classic example of a market failure, where the value to society of ULCATS is tremendously high, but much of that value will be in the long-term, and most of the risks are in the near-term.

Some argue that high flight rates will lower launch costs for expendables just as much as it does for reusables. Do you disagree?

It is in the economic interest of some to perpetuate the status quo. Yes, at least one expendable launch vehicle (ELV) company argues the only thing that matters is flight rate, and that the cost of ELVs will go down essentially the same amount as the cost of RLVs for any given flight rate. This argument is demonstrably false. If their argument were true, we would use expendable airplanes today.

Their costing methodology ignores substantial cost savings in operating costs that can be achieved with RLVs, but not with ELVs. The first time an RLV is integrated and tested, the costs are similar to integrating and testing an ELV. The same high integration and test costs exist in the first acceptance flight of every airplane. However, this is where the ELV cost model breaks down. We build airplanes with ever-higher reliability to support ever-higher flight rates in order to lower operations costs for the flights that come after the acceptance testing. True reusability leads to much higher levels of reliability, which leads to a transformation in operations costs. After an airplane completes acceptance testing, its reliability to do the job in its flight regime is validated. Expensive tests, maintenance and inspections are no longer required between every airplane flight. Instead, we have automated health maintenance systems that watch many indicators for signs of wear and tear. This allows the operator to fuel up, and rapidly fly again, with minimal labor (operational) costs between flights. The same "aircraft-like operability" benefits will be gained by RLVs as well, and this is where most of the cost savings will be found that lead to ULCATS.

If the USG is not willing to commit in advance to buying services, why do we think industry will partner?

Many of the companies are investing right now in systems leading to ULCATS for commercial purposes. We are proposing to accelerate their existing long-term plans, and to improve the likelihood they will successfully develop these systems. More specifically, multiple companies have told us they are willing to partner right now with the USG, even without a commitment in advance to buy their services.

If industry will partner for development without early USG commitments to buy services to drive up demand, isn't this a good deal for the USG? Why should the USG commit to early purchases of services?

If only one company develops ULCATS systems, in existing markets they will only lower prices enough to drive other launch companies out of the business, and then they will raise prices. If two companies invest, the competition will drive down prices, but they will likely segment the market based on the differences in capabilities of the systems, and the price competition might be minimal.

With three companies, the competition will be much more aggressive and price reductions more significant. The companies will understand this in advance and factor this into their investment decisions. If they do not perceive the market growth to be large enough, it may be difficult to persuade three companies to seriously invest in ULCATS, even with the USG as a partner.

"Program X" used an OTA agreement and failed, so why will OTAs work here?

In the last 30 years, the only new rockets successfully developed in America have used OTAs. Traditional rocket development methods were tried many times in that same time period and failed. Other OTA "lessons learned" research has shown OTAs are most effective when executed by a dedicated team that understands the technology, innovative development processes, and is unafraid of discarding "business as usual" techniques. Project teams that use an OTA, but impose the same traditional processes, procedures, and methods, will have traditional results. Some OTA projects have failed for reasons having nothing to do with the fact that it was an OTA.

Why Should the USG get involved in a private US industry decision?

The implications of ULCATS have tremendous near-term and long-term implications to US national security, which means that it is too risky to leave this solely to the decisions of private markets. Further, the USG is already tremendously involved in all space development projects and activities. In the first 50 years of the Space Age, the USG has completely dominated decisions about investment and operations in space. On the spectrum of USG actions, this proposal is consistent with recent trends in the transition of transferring more responsibility for development and operations to US private industry, which some call privatization.

Appendix B — Purpose-Built Organization Implementation Plan

Appendix C — Detailed Recommendations for Proactive Approach to Interagency and National Policy Shaping

National Security Strategy & Policy

Based on Finding F.3.1 through F.3.3 starting on page 29, we provide the following recommendation:

RECOMMENDATION [R.6]: American leadership is vital to creating international space law based on Western values: The United States needs to lead in utilizing space and space resources, in order to establish common international operating principles based on Western values and law. Ceding the "high ground" to other nations to develop this nascent area of the law would enable precedents that directly conflict with Western beliefs and ethics, which could have devastating long-term repercussions to national security, economic policy, and fundamental American freedoms and liberties.

National Leadership

Based on Findings F.3.4 and F.3.5 on page 32 we provide the following recommendation:

RECOMMENDATION [R.7]: DOD should propose that the White House create an interagency organization resident in and staffed by the Executive Office of the President to oversee ULCATS-related investments and reforms across space-involved agencies: USAF should make the case that, unlike typical policy coordination groups, this organization should have an active management role in ensuring that agency actions are fully aligned with the national goal of achieving ULCATS. It may be possible to use existing statutory authority for the National Space Council, under the Vice President, to meet this need. However, lessons learned from previous RLV-related projects demonstrate that an ULCATS initiative will be seen as disruptive to the status quo and will be resisted — ULCATS will need the active support and protection of national leadership. It should be the first initiative on the Council's agenda, and the Vice President needs to be personally involved in overseeing its progress.

Commercial Policy & Regulations

Based on Finding F.5 on page C-3, we provide the following recommendations:

RECOMMENDATION [R.8.1]: Begin the transition of public safety at launch ranges to civil agency: It is unlikely, if not politically impossible, for the USAF to invest the amount of resources needed in the federal ranges to support the commercial launch rates that are projected in the future. Responsibility for public safety for non-DOD users should be transitioned to a civil agency, which may only oversee privately-conducted flight safety approaches. This will allow DOD to focus its resources on actual military tests and operations requiring the range, potentially including current range public safety systems. The USAF and other DOD range users would be "first among equal" as range tenants, with a preemptive right to additional capacity for urgent national security purposes.

RECOMMENDATION [R.8.2]: Create a commercial "airport-like" operating environment at federal ranges by transferring responsibility for overall operation of the federal ranges, starting with support infrastructure and services, to an independent

spaceport authority: USAF should initiate a methodical transfer of federal launch range support infrastructure (and its maintenance/improvement) and related services to a more appropriate non-federal entity, and eventually overall operational responsibility for the ranges, while retaining title and any facilities/infrastructure needed to support USAF mission requirements.

RECOMMENDATION [R.8.3]: Restructure commercial launch licensing: For 30+ years Congress has enacted strong, enabling legislation that allows the Department of Transportation (DOT) to minimally regulate and actively promote the US space launch industry. The law is currently implemented by a relatively small organization buried within the much larger FAA. Achieving ULCATS will require a radical change in how DOT carries out their existing mandate or the launch licensing process will slow down, if not in fact prevent, achieving the frequency and responsiveness of launches that will typify ULCATS. USAF can use Recommendation 3.1 as good faith leverage to spur DOT action on this recommendation.

RECOMMENDATION [R.8.4]: Restructure spacecraft licensing: NOAA requires a modernized law and book of regulations for commercial remote sensing, and the FCC needs to better allocate frequencies and license spacecraft constellations. DOD is an active participant in interagency reviews for both NOAA and FCC, and therefore has some entre to promote reforms on a peer agency basis.

RECOMMENDATION [R.8.5]: Improve collision models: The Air Force should expend the resources necessary to increase the reliability of the current orbital collision models based on the new data provided by the "Space Fence".

RECOMMENDATION [R.8.6]: Research debris remediation technologies: The DOD and NASA should support research into orbital debris remediation technologies. Because "one man's debris removal device is another man's ASAT," the commercial space sector should operate these systems, under license from a civil space agency, and with input from the USAF to protect national security space assets.

Commercial Regulation & Policy

SUMMARY: All paths to a future with ULCATS require significant changes to commercial regulations and policies, as current policy and regulations were designed in a world that assumed expensive and infrequent launch.

ULCATS will be the result of significant advances in the affordability, quantity, reliability, and capability of space transportation goods and services designed, manufactured, and operated by US commercial providers. A global market of private and public customers will consume these ULCATS goods and services. All paths to this future require an enabling legal, regulatory, and policy environment as critical "business infrastructure" within which the ULCATS marketplace can function.

FINDING [F.5]: Regulatory systems – and some underlying laws – for space launch were not designed to support ULCATS and need to be redesigned: While public law for commercial space transportation has evolved beneficially over 3 decades, the regulations which implement it have not, and lag far behind technological and business developments. As such, the government organizations and processes that deal with launch providers are still largely aligned with launch being a high-cost, infrequent, risky, and very specialized activity. However, the law as well as regulations for customers of commercial space launch are much less progressive, and will restrict potential increases in demand that would reinforce improvements in space transportation supply.

The legal and regulatory systems, processes, and institutions which make up this infrastructure are already showing clear signs that they cannot support the current growth trend in frequency of launch or spacecraft activity. Examples include:

- How the Federal Aviation Administration (FAA) licenses launches and reentries,
- How National Oceanic and Atmospheric Administration (NOAA) licenses remote sensing spacecraft and ground systems and restricts their business activities,
- How the Federal Communications Commission (FCC) allocates frequencies and licenses communications for all space systems,
- How the USG enforces orbital debris mitigation, tracks orbital conjunctions, and conducts research into orbital debris remediation technologies, and
- How the federal ranges, particularly the Eastern Range, operate to protect the public safety during launches, and how they operate as hosts/providers of space transportation infrastructure and related services.

These regulatory bottlenecks may be solvable over the very short term by applying more money to the various agency programs, but no amount of money can add sufficient capacity to these systems to support the volume of activity expected, and the responsiveness required, of ULCATS. Any significant increase in flight rate from the status quo trend (50% or more) will require significant, if not radical, change in the various licensing and operational government organizations.

Finding F.5 directly supports Recommendations R.8.1 to R.8.6 located on pages C-2 and C-3.

Appendix D — Findings, Recommendations and Observations Table

Table 1.0 — Findings (F), Recommendations (R), and Observations (O)			
Partnerships			
Finding No#	**Finding**	**Recommendation No#**	**Recommendation**
F.1.1a Partnership	The capabilities provided by ULCATS/Fast Space could enable the Air Force to leverage emerging commercial technologies and investments, thereby impose significant costs on adversaries across all five of its core missions through operational agility	R.1 Partnership	Partner with US commercial firms pursuing ultra low cost access to space (ULCATS) using the DOD's Other Transaction Authority
F.1.1b Partnership	Whole new architectures and concepts that will transform USAF dynamic C2 and ISR will become economically affordable and technically feasible with ULCATS.		
F.1.1c Partnership	Commercial RLVs with rapid turn-around could provide the United States a prompt global strike capability		
F.1.1d Partnership	An early success may be theater pop-up missions for suborbital RLVs		
F.1.2 Partnership	There is a strategic common ground between rapid developments in private space industry and USAF needs		
F.1.3 Partnership	ULCATS could provide a Stabilizing Deterrent to War		
F.1.4	While Air Force needs for space operations that are "aviation like" are aligned with commercial interests, they are not identical.		
F.1.5 Partnership	Industry Believes RLVs Technically Achievable based on Today's Technology		
F.1.6 Partnership	Commercial Methods are More Economically Affordable for Developing RLVs		
F.1.7 Partnership	Primary Barrier to RLV Development is Commercial Business Case		

Finding No#	Finding	Recommendation No#	Recommendation
F.1.8 Partnership	Industry Supports Use of Risk-Sharing OTAs for Commercial Partnerships		
F.1.9 Partnership	1st Generation Commercial fully Reusable LVs are Technically Feasible		
F.1.10 Partnership	Full Reusability is Still a Difficult Technical Challenge		
F.1.13 Partnership	The development of multiple competing fully-reusable launch vehicles could lower prices by a factor of 3X in the near-term		
F.1.14 Partnership	Competition is Required for Large Reductions in Price in Commercial Markets		
F.1.15 Partnership	Advances in Reusability Likely to Start Virtuous Cycle		
F.1.16 Partnership	In the Long-Term, a Launch Price Reduction of 10X is achievable		
F.1.17 Partnership	Big LEO Constellations Unlikely to Close Business Case in Near-Term without RLVs		
F.1.18 Partnership	Current Launch Demand Not Large Enough to Support a 10X reduction (ULCATS)		
F.1.19 Partnership	New Markets Enabled by 3X Cost Reduction Can Lead to ULCATS		
F.1.20 Partnership	OTA-based Commercial Risk-Sharing Partnerships are Successful		
F.1.21 Partnership	Commercial OTA Partnerships are much lower cost than Traditional FAR-Based Methods		
F.1.22 Partnership	The DOD has the OTA Authority (10 USC 2371b) it Needs for ULCATS Partnerships		
F.1.23 Partnership	A portfolio approach to commercial partnerships, on both demand and supply sides, will lower overall strategic program risk and has the highest likelihood of success		
Joint Requirements			
Finding No#	Finding	Recommendation No#	Recommendation

Finding No#	Finding	Recommendation No#	Recommendation
F.2 JROC Requirements	Current JROC requirements are not enabling to Fast Space	R.2 JROC Requirements	Integrate consideration of Fast Space and RLVs into the Joint requirements and acquisition process
Shape Interagency Environment and National Policy			
Finding No#	**Finding**	**Recommendation No#**	**Recommendation**
F.3.1 National Security Policy	International space law, including the Outer Space Treaty of 1967, is flexible enough to allow for military usage of outer space short of deploying nuclear weapons and weapons of mass destruction or actively interfering with the activities of others	R.3 Shape Interagency Environment	Shape the interagency environment. To maximize the operational and strategic utility of ULCATS, the SECAF should adopt a proactive approach to help shape national policy
F.3.2 National Security Policy	Because much of international space law is based on customary international law, whoever leads in space development is likely to establish operational legal principles, values and practices on the space frontier that will last far into the human future		
F.3.3 National Security Policy	It is the national security interests of the United States to establish the Western principles of free trade and commerce, including free enterprise development and use of space resources, in international common law.		
F.3.4	Achieving ULCATS will require national-level leadership beyond DOD		
F3.5	ULCATS will create economic and diplomatic soft power benefits, which also contribute to US national security		
Purpose-Built Organization			
Finding No#	**Finding**	**Recommendation No#**	**Recommendation**
F.4 Purpose-built Organization	Traditional USG acquisition methods, or a traditional operationally focused acquisition office, are likely to fail at effectively partnering with commercial space industry	R.4 Purpose Built Organization	Create a purpose-built organization to manage commercial ULCATS efforts

Technology				
Finding No#	Finding	Recommendation No#	Recommendation
F.1.11 Technology	New Technology Required to Achieve Continuous Improvements in Aircraft-like Operations	R.5.1 Technology	Develop Commercial-Industry Prioritized Investment List of Next Generation RLV Technologies
F.1.12 Technology	The Systems Integrators are in the best position to Lead the Technology Prioritization Process		
		R.5.2 Technology	NASA and AFRL should Invest according to Commercial Industry's Prioritized List of RLV Technologies

National Security Policy				
Finding No#	Finding	Recommendation No#	Recommendation
F.3.1 National Security Policy	International space law, including the Outer Space Treaty of 1967, is flexible enough to allow for military usage of outer space short of deploying nuclear weapons and weapons of mass destruction or actively interfering with the activities of others	R.6 National Security Policy	American Leadership is vital to creating international space law based on western values
F.3.2 National Security Policy	Because much of international space law is based on customary international law, whoever leads in space development is likely to establish operational legal principles, values and practices on the space frontier that will last far into the human future		
F.3.3 National Security Policy	It is the national security interests of the United States to establish the Western principles of free trade and commerce, including free enterprise development and use of space resources, in international common law.		

National Leadership				
Finding No#	Finding	Recommendation No#	Recommendation
F.3.4 National Leadership	Achieving ULCATS will require national-level leadership beyond DOD	R.7 National Leadership	Propose that the White House National Space Council oversee ULCATS-related

| F.3.5 National Leadership | ULCATS will create economic and diplomatic soft power benefits, which also contribute to US national security | | reforms across space-involved agencies |

Legal & Regulatory			
Finding No#	Finding	Recommendation No#	Recommendation
F.5 Commercial Regulation & Policy	Regulatory systems — and some underlying laws — for space launch were not designed to support ULCATS and Need to be Redesigned	R.8.1 Regulatory	Begin Transition of Public Safety at Launch Ranges to Civil Agency
		R.8.2 Regulatory	Create commercial "airport-like" operating environment at federal ranges by transferring responsibilities to independent spaceport authorities
		R.8.3 Regulatory	Restructure Commercial Launch Licensing
		R.8.4 Regulatory	Restructure Spacecraft Licensing
		R.8.5 Regulatory	Collision Models Need to Be Improved
		R.8.6 Regulatory	Research Into Debris Remediation Technologies

OBSERVATIONS	
Observation No#	Observation
O.1	The capabilities provided by RLVs do not neatly align in any one Core Function Lead Integrator (CFLI)
O.2	An initiative to develop much lower cost launch vehicles is not sufficient, on its own, to deliver a 10X Reduction (ULCATS).
O.3	Key drivers of ULCATS are: (a) Flight Rate, (b) Reuse Rate, and (c) the Labor Intensity of Operations, and each of these factors are inter-related.
O.4	Labor Intensity of Operations is the Most Ignored Key Cost Driver.
O.5	Expendables are Unlikely to achieve ULCATS.
O.6	Traditional Investors are not likely sources of capital for ULCATS.
O.7	Philanthrocapitalists and Strategic Investors will lead.

Appendix E — Assessment Team

Primary Assessment Team

Charles Miller, NexGen Space LLC, *Principal Investigator*
Craig Leavitt, Col., USAF (ret.), National Defense University, *Lead Strategist*
Ben Muniz, NexGen Space, *Project Manager*
Margo Deckard, NexGen Space, *Deputy Project Manager*
Tim Huddleston, NexGen Space LLC
Tshela Mason, NexGen Space LLC
Tyghe Speidel, NexGen Space LLC
Lt. Col. Thomas Schilling, Air University
Lt. Col. Peter Garretson, Air University
Dr. M.V. "Coyote" Smith, Col., USAF (ret.), Air University
Harry Foster, Col., USAF (ret.), Air University
Col. Jeffrey Donnithorne, Air University
Dr. Alan Wilhite, Wilhite Consulting LLC
Hoyt Davidson, Near Earth LLC
Joshua Hartman, Renaissance Strategic Advisors
Bill Bruner, Col., USAF (ret.), New Frontier Aerospace
James A. M. Muncy, PoliSpace
James Dunstan, Mobius Legal Group
James Ball, Spaceport Strategies LLC
Alan Lindenmoyer, Lindenmoyer Aerospace Services LLC
Richard L. Dunn
Dr. Joel Sercel, ICS Associates
Leon McKinney, McKinney Associates
Edgar Zapata, NASA Kennedy Space Center
Roger Lepsch, NASA Langley Research Center
Dr. Daniel Rasky, NASA Ames Research Center
Dr. Paul Jaffe, Naval Research Laboratory
AU Faculty and Students

Independent Technical Review Team

Dr. Douglas Stanley (Chair), National Institute for Aerospace
Jeff Thornburg, Interstellar Technologies, LLC
Jay Penn, Aerospace Corp

Independent Senior Review Team

N. Wayne Hale, Jr., Special Aerospace Services (Chair)
Kent Joosten, Onyx Aerospace
Howard "Mitch" Mitchell, Major Gen., USAF (ret.), Aerospace Corp
Robert Mitchell, Robert A.K. Mitchell Consulting
Gary Payton, Professor, USAF Academy, Col., USAF (ret)

Appendix F — Acronyms

A2/AD	Anti Access/Area Denial
ACC	Air Combat Command
AFRL	Air Force Research Laboratory
AFFOC	Air Force Future Operating Concept
AU	Air University
C2	Command and Control
CFLI	Core Function Lead Integrator
COCOM	Combatant Command
COTS	Commercial Orbital Transport Services
CPGS	Conventional Prompt Global Strike
DARPA	Defense Advanced Research Projects Agency
DOD	Department of Defense
DOT	Department of Transportation
DDT&E	Design, Development, Test and Evaluation
EELV	Evolved Expendable Launch Vehicle
FAA	Federal Aviation Administration
FAR	Federal Acquisition Regulations
FCC	Federal Communications Commission
GDP	Gross Domestic Product
GEO	Geosynchronous Earth Orbit
GIISR	Global Integrated Intelligence, Surveillance, and Reconnaissance
IR&D	Internal Research & Development
ISR	Intelligence, Surveillance, and Reconnaissance
LEO	Low Earth Orbit
LV	Launch Vehicle
MNS	Mission Needs Statement
MSP	Military Space Plane
NAI	National Aerospace Initiative
NAFCOM	NASA Air Force Cost Model
NASP	National AeroSpace Plane
NASA	National Aeronautics and Space Administration
NSDO	NewSpace Development Office
NDU	National Defense University
NOAA	National Oceanic and Atmospheric Administration
NRC	Non-Recurring Cost
OODA	Observe Orient Decide Act
OSD	Office of Secretary of Defense
OT	Other Transactions
OTA	Other Transaction Authority
PDSA	Principal DOD Space Advisor
PNT	Precision Navigation and Timing
PPP	Public Private Partnership
RDT&E	Research, Develoment, Test and Evaluation
RFI	Request For Information

RLV	Reusable Launch Vehicle
ROI	Return on Investment
SECAF	Secretary of US Air Force
SECDEF	Secretary of Defense
SME	Subject Matter Expert
SSA	Space Situational Awareness
SSP	Space Solar Power
TSTO	Two-Stage-To-Orbit
US	United States
USAF	United States Air Force
USG	United States Government
VTVL	Vertical Take-off and Vertical Landing
WWII	World War 2
XS-1	Experimental Spaceplane-1

www.ingramcontent.com/pod-product-compliance
Lightning Source LLC
Chambersburg PA
CBHW080553170426
43195CB00016B/2781